Memories of a SUITCASE FARMER

Family Farm to City Boardroom—
The Country Stays in the Boy

DAN DANIELSON

 Published in Canada by Dan Danielson with assistance by Articulate Eye.

Cover, interior design, and scans by Articulate Eye, Saskatoon

Cover photograph by *The StarPhoenix*

Edited by Sharon Danielson

Proofread by Harold Pexa

Printed in Canada

Library and Archives Canada Cataloguing in Publication

Danielson, Dan, 1949-, author

Memories of a suitcase farmer : family farm to city boardroom—the country stays in the boy / Dan Danielson.

ISBN 978-0-9947444-0-1 (pbk.)

1. Danielson, Dan, 1949-. 2. Farm life--Saskatchewan—Tyner—History. 3. Agriculture and state—Saskatchewan. 4. Agriculture and politics—Saskatchewan. 5. Farmers—Saskatchewan—Biography. 6. Saskatchewan—Biography. I. Title.

S522.C3D35 2015 630.92 C2015-902527-3

Publisher:
Dan Danielson
danielson@sasktel.net

CONTENTS

This book is dedicated to my father and mother, Urvan and Geneva (Thompson) Danielson who came before me and provided great love, support and guidance so I never felt poor even though as a family we had limited material wealth. Also to my best friend and wife Sharon (Oliver) Danielson who supported me and worked hard both on and off the farm to guarantee our success. Finally to our sons Danton and Brandon who were with us on much of the journey and have provided additional meaning to our lives.

To Les

"We can all make a difference"

Dan R Danielson

one

INTRODUCTION

A few weeks ago, Sharon, my first and only wife and I made a rushed trip out to our family farm, which for the most part is now operated by our neighbour Cory and his family. We still perform some of the duties on the land that we continue to rent from others, but Cory and his family handle most of the work. Of the thirteen quarters that make up our farm, Cory cash rents six from us and we contract him to custom work the remaining seven quarters. In the first instance he takes all the risk, and in the second instance we take the risk in terms of crop production and the prices but we get paid for the crop. Not unlike many other family farms in Saskatchewan, our farm has become part of a much larger farm operation, in this case Cory's farm enterprise, three or four times our farm size.

Every time we go to our farm where we keep our house and yard maintained, we pass within one mile of my once active and friendly hometown of Tyner. We look east and say to each other, "There is where Tyner used to be." On this most recent trip, Sharon and I decided to pop into that little town to have a look around and see what was up. As we came closer and closer to town, my mind began to race backwards with enthusiasm as I quickly processed multiple memories from my past. Unfortunately, I already knew quite well the brutal fate of my town, which had virtually disappeared. However, as one enters the perimeter of a place where you spent so many delightful years of your life, it is impossible to suppress the intrinsic message of sadness that burns into your mind. The school is gone, the hall is gone, the two stores are gone, the post office is gone, the skating and curling rinks are gone, the railway station is gone, the three elevators are gone, and the railroad track has been pulled up. Even the "book-end farmyards" at each end of our town have disappeared into flat grassland. "Where has my hometown gone?" I quietly and sadly asked myself. With justification I could not resist the temptation to lament for my past.

Ironically, today there is only one family of three people who live in Tyner, and they only recently moved in from Calgary, after our town had once become completely vacant. The new residents, a family of a middle-aged mom and dad with a young adult son, told us they saw an ad on the internet for this one last remaining home in Tyner so they bought it, as a way to leave a too big city and go to a quiet farm town. These people made us truly believe that they came here for solitude. The mom was more than willing to talk to us for an extended time out in her yard. She seemed anxious to find out what I knew about her new hometown when she learned that it was my old hometown. With nostalgic and romantic memories bouncing like ping-pong balls in my head, I was happy to reflect on my past for her information and for my nostalgic pleasure.

Sharon and I later mused while we were on the road home that this potentially lonesome family had gone from too big and noisy, to too small and quiet. Even so, when we later returned to our place in Saskatoon, my mind was still in a glow with thousands of stimulating thoughts going back to how my life evolved during and after my time in Tyner. My collection of memories could be compared to rocks on the field; when you picked one up and put it on the wagon or stone boat, almost immediately another would appear. Ironically, as I thought more and more about my little hometown, it dawned on me that while we had a hometown proper in Tyner, we didn't have a single town which offered all the services that were required even back then. The real truth for most small towns is that they were part of a constellation of towns, working and playing together to collectively provide as many social or business services as they could. It was at that very moment I decided I would write down as much as I could remember about my life in these rural communities, starting from the time my family moved to the town of Tyner for me to start school at age six in 1955.

Our family farm at that time was about six miles away from Tyner, but my dad and mom decided they could farm from town in order to be closer to the school. It was the school where my mom had once been a teacher before my dad, a smashing young WWII veteran, courted her to the farm and into farm life. I grew up in that small town until I was about eighteen years of age, and then went to Saskatoon for university in the winter months, returning home to work on my parents' and our neighbour's farm in the summer. After graduating from university, I accepted several different employment opportunities in Saskatoon or Regina, often just for the winters as I went home to grow my own farm operation.

Most of the farms in our district were straight grain operations, which meant that without livestock our winters were relatively free of most farm duties, so folks were able to seek off-farm work or education opportunities

The Sam Danielson Homestead (Grandpa)—my first home

My Mom as a teacher in Tyner

My Dad in the army

during the winter. Those opportunities were another form of farm diversification often needed to enhance the farm income. My life often meant living and working on the farm in the summer and then moving or commuting to other locations in the winters to subsidize my farm income, to improve my education, or to connect with other services or experiences. Sometimes I would take full-time year round employment and do the farming on weekends and holidays, as hectic as that could be.

The term "Suitcase Farmer" accurately describes my nomadic lifestyle. In order to fit everything into a busy life with my diverse career and family experiences, it would not be unusual to find my heavy old Samsonite suitcase—the one my wife backed over with the truck—partially packed near my bed, ready to be filled to the top and forcefully closed en route to a different location close to another job, the university, or a boardroom meeting. That spirit of mobility was best illustrated later in my career, when I would get off my tractor in the field on our farm about eleven o'clock in the morning, then travel into Saskatoon to chair the largest city school division in the province that afternoon and evening.

Now I want to invite you to consider taking an informal literary journey with me, as I try to tell you about my love for our province, our people, our family farms and our rural communities as told from my memories as a "Suitcase Farmer." I ask you to follow with me, as one by one I begin to empty my memory box located in the very back corner of my fragile mind. I hope we can take this journey for your enjoyment and for my own nostalgic pleasure, because I want to tell my story before it is forgotten.

Perhaps I should present a few disclaimers for the sake of accuracy, and to protect my creative freedom. Please keep in mind that when you get to my age two things start to go bad for you; one is your memory, and I can't remember the other one. Therefore, the facts and events in this story will be as complete as I recall, but may be unintentionally different from what others remember. I like to say that I might be wrong but I won't be wrong for long, so I am more than willing to accept validated corrections. Also many of the names have been changed to protect the guilty, and some degree of embellishment and creative license has been included for the sake of interest. Having said that, much of this writing is guided by the worldly principle that truth is stranger than fiction. Since I turned sixty I feel compelled to offer free advice, but my kids would say it started way before that, and with a teacher background undoubtedly I will either knowingly or unconsciously insert a few lessons into my story. I promise to limit my teaching and preaching as much as I can.

My life has not been all work and I've had at least my fair share of fun doing more than enough crazy things. I played several sports and rarely found a party that I didn't enjoy. I want to tell you about the fun we had at the Harvest Moon Ball, the White Bear Pub, the Chinese Cafés, or the School Graduation Picnic. There is also one leisure activity which has, since I was about twelve years of age, given me great pleasure and much relaxation while connecting me with mother nature and expanding my respect for our environment. That activity is the annual "Goose Hunt," which for as long as I can remember has been a local activity in my home area. Even though most years, when late autumn arrives I have been away pursuing my complementary career objectives, I still strive to be home for at least a few days for a goose shoot. Any story close to my heart must include the simplicity and beauty of nature, the cheers of friendship, the reward of physical exercise, and the pride of ownership gained from "The Goose Hunt" on our own land at "Goose River Lodge."

Not so long ago on one of my ego trips I asked my wife Sharon if, in her wildest dreams, she ever thought that a farm boy from Tyner, Saskatchewan could go on to university and earn three degrees including a Masters Degree in administration and take on interesting employment opportunities with the Saskatchewan Land Bank, Saskatchewan Farm Ownership Board, the Federal-Provincial Partnership Agreement on Rural Economic Development, or teach school in a northern community. Further, he would get elected trustee and board chair of the Saskatoon Public School Board, the largest school division in the province, and be elected as an executive member of the Saskatchewan School Boards Association (SSBA), as well as be appointed provincial education director on the Saskatchewan Assessment Management Agency (SAMA), then also be elected as a Director of the first Saskatchewan Wheat Development Commission (SWDC). In addition, he would also build a successful family farm. All she said in response was: "Daniel, you are not in my wildest dreams." She always has a way of keeping me humble.

On that note I can only say: "From the family farm to the city boardroom, the country stays in the boy. As a result, there are many human interest stories to be reported by a 'Suitcase Farmer.' Should we begin?"

two

WHERE HAS MY HOME TOWN GONE?

Local folklore, as I hear it told, says that some of the gentle ladies of our proud community thought it was high time that I should get off to school in order to do some book learning, and more importantly learn some refined language skills. That opinion, I was told many years later, resulted from my unceremonious outburst while reporting what had happened when I was helping my dad at the morning milking chores with "Old Gluepot." She was our usually trustworthy, reddish-tan, Hereford milk cow with large brown eyes that usually prompted your affection.

It is a long story, but as I was informed, our productive old milk cow had injured her udder area by accidentally catching herself on some exposed barbed wire ends, where my dad had fixed the surrounding four-wire fence. Soon after that accident, when she was still in much discomfort, my dad and I came into her milking stall, inside of our conventional white cedar sided, hip roofed barn with the fancy dark green trim. Old Gluepot was moving into her regular position for the milking process when she bumped her tender self on the stall gate. In some considerable pain, she jumped back and firmly kicked my dad as he innocently stood behind her, landing him flat on the floor, face down in some mushy cow manure. Fortunately, the soft and deep flax straw used for bedding, had protected him from much physical injury, but the surprise and embarrassment obviously hurt him more. Now, my dad was normally a mild-mannered, quiet-spoken, gentle farmer. He

"Old Gluepot"—our milk cow

even considered our old cow as his devoted friend. However, that day she must have temporarily forgotten how my dad, with me at his side, had personally carried her new reddish-brown newborn calf with a white face and large white side patches, out of the blizzard-like snowstorm from the pasture into the cozy warm barn. Things were different that day under those violent and demeaning circumstances, even with the usually comforting sweet scent of fresh alfalfa bales. My dad jumped up off the floor and clearly addressed that newly wild milking cow, in language that would make a cowboy blush. He firmly hit our milk cow with the handle of a nearby pitchfork he had quickly taken from its place on the unpainted board wall.

Now that may have been the end of the story at any other time, except for what I did next. That fateful event happened on one of those special days when several of the most sophisticated, proper, and dignified ladies from around the countryside, were at our farm, because my mom was the talented and valued local hair stylist. All was well, I was told, until I came into the farm house at five years of age, and like a miniature news anchor succinctly reported word for word with emphasis and expression, what my temporarily angry father had communicated in no uncertain terms to "Old Gluepot." In my replay monologue, described by others as a profane and animated rant, I also managed to knock over a double layer chocolate cake with white icing, that would have been part of the day's final festivities for those pampered ladies, had it survived. As important as I thought my story was, my performance was to meet with mixed reviews. I guess most thought it was funny, but some didn't. Obviously, they concluded, it was time for me to go to finishing school, so off we went to Tyner, located five miles east in what was called the "Tyner Flat." The original Gravelridge School within a mile of our farm had been closed.

Fortunately I guess, by grade five, I had already become quite fluent with the language and could even add a little "spin" as I told my class about the failures of our old green Chevrolet truck. My analysis of a bad mechanical situation recounted how the spark plugs wouldn't spark, the flywheel couldn't fly, the crankshaft wouldn't crank and the pistons wouldn't work either. Mr. Mallett, my school principal at the time, told me many years later what I had said because he said it amused him and the class.

Gravelridge School location—where my Dad went to school—on land we now own and farm

Our Hometown of Tyner

I grew up in the small town of Tyner, Saskatchewan not "Tiny," Saskatchewan as it was often mistakenly called, much to my displeasure. Tyner came into existence in the early nineteen twenties along with many other prairie towns placed about eight miles apart along a CNR rail line, which served as the lifeline threading together numerous small rural communities like pearls of the prairies. Someone once speculated that God gave us all these small towns, each with three or four eighty-foot high grain elevators, as a helpful start on a shelterbelt, needed to protect us from the fierce prairie winds. Story has it that Tyner got its name simply from the fact that a railroad manager at the time of our town's origin had that name. Tyner is located along the CNR Branch line heading southeast from Eston to White Bear. Originally that line served a string of busy places with grain elevators and business outlets including Eston, Whitley, Isham, Tyner, Lacadena, then over a large wooden bridge through a deep valley to White Bear at the end of the line.

Tyner is about two hundred kilometres southwest from Saskatoon. You head out on highway seven through Rosetown, then due south almost to Elrose, where you turn west again toward Eston. Within about twenty kilometres, when you see Plato to your right, you turn south again to your left, arriving at Tyner in about seventeen clicks. As you come south down that high Plato Hill, as we call it, you begin to see a giant flat prairie grid, consisting of rectangular green or golden fields of crops or dark summer fallow fields. That area is known as the Tyner Flat. It is so flat that water drainage is difficult, resulting in a patchwork of quite shallow but extensive sloughs scattered throughout the fields.

The Tyner Valley Acquifer is a major underground drainage channel running from the South Saskatchewan River near the Saskatchewan–Alberta border past Tyner northeast past Saskatoon where it meets the Battleford Valley Acquifer and together they feed into the North Saskatchewan River. It is about one hundred meters below the surface and is about sixty to ninety meters thick and about six-thousand to ten-thousand meters wide. This acquifer receives flow from the South Saskatchewan River and also vertical flow from overlying glacial drift acquifers.

Where the Deer and Antelope Play . . .

Pronghorn Antelope *(photo courtesy of Glenda Rossignol)*

Mule Deer Buck *(photo courtesy of Glenda Rossignol)*

In the early spring and fall the sky will be alive with the sound of music, as you are greeted by the songs of game birds, including ducks, cranes and geese. Almost anywhere across our local prairie landscape, you might see several stately and very fast pronghorn antelope sporting their light tan coats while calmly grazing on the land. Also, above the tall summer grain crops, you can expect to see the prominent antlers of large grey mule deer sticking up like branches of trees that have lost their leaves. In the early summer those antlers are covered in velvet skin, and have not yet hardened into the mature hard fighting horns for the fall rut. White Tail or Flag Tail Deer can also be observed scattered near any of the scarce clumps of bush or areas of tall grass. Foxes, coyotes, skunks, porcupines and rabbits help complete the natural cast of characters who take their place in this flat prairie environment. Our area offers you very few trees or natural vegetation, other than a few hand-planted trees or bushes in the sparsely scattered farmyards, or in the town of Tyner itself.

I lived for twelve years in two different Tyner homes with my parents, and later three other houses for about six years as a young bachelor farmer, before my wife Sharon and I established our new farmyard six miles southwest of town in 1980. When my family and I arrived in Tyner in the fall of 1955, there were about twenty-five occupied homes and the local businesses served about twenty-five nearby farms.

Our Stores and Post Office

I clearly remember a large and busy grocery store, along with an equally large handy hardware and dry goods store, joined together by the attached, low profile post office. I still vividly remember Friday night as train night, when people from town and country gathered on Main-Street. Cars and trucks angle parked in an organized row along the wide main wooden sidewalk that shook and squeaked as busy shoppers hustled to and fro in front of the stores. The parked cars always left space by the red or blue gas pumps with the large handles on the side, located right in front of the hardware store.

Friday night, as I mentioned, was train night when the good old coal burning, steam-powered, hard black iron horse delivered the mail, grain

The Hardware Store—the Post Office—the Grocery Store

cars for the elevators and other freight necessities. Later I recall that steam engine was replaced by a more powerful diesel locomotive. The high silver cream cans with the farmer's name boldly written on them, full of milk products from the cattle farms, would be placed on the station platform by the track ready for a trip to Saskatoon. The cheques for last week's shipment should arrive in the large mailbags carried off the train over to the post office. Many farmers, usually the farm wife, waited patiently for those "cream cheques." which sometimes made up a significant part of the family budget.

During the early weekdays, Black's Transport from Swift Current and Turner's Transport from Kyle would bring in the household and food supplies to be sold in the stores apart from items purchased directly from the farm gates. Perhaps it was more complicated than I remember it to be, but in my mind the work necessary to run that grocery store appeared very clean, simple and routine to me at the time. It seemed like a great place to work and visit leisurely with customers, protected from the harsh prairie weather that was part of normal farm life. To me it was amazing how so many groceries or household items of all kinds could be tallied through that noisy cash register, bagged in heavy brown paper bags, and lugged out by the customer. Sometimes, if not so busy, the storekeeper with his or her full-length green apron would carry out your purchases. I especially loved watching the slicing and wrapping of that giant, round, yellow block of cheddar cheese, sitting on "cheese cloth" and covered with a transparent lid, on the table right by the counter. That was a pleasure we all enjoyed, especially when a chunk of it was destined for our fridge at home, to be later placed between two generously buttered soda crackers, when my family and I had our precious bed lunch and a special time together.

Managing the hardware and dry goods store, I felt, was a much more complicated and scientific matter. When I think back, I remind myself how that job must have been an art as much as a science, to know how many pairs of rubber boots, pairs of gloves, decks of playing cards, mouse traps, hockey sticks, rolls of tape, nails or screws etc. to stock on those numerous shelves. Then there were the sideline chores to sell fuel from the gas pumps out front, or sell iron from the rack at the back, and leather or wood supplies from the attached warehouse at the side. I remember how one energetic family would operate both stores and the post office, doing all those duties.

As cars and trucks were modernized and transportation became easier, it became more difficult to make a go of it running these local family businesses. Most local folks knew that because of economies of scale and lower freight costs, better prices could be found at larger towns further away, so some people would stray away with their business. However, in order to retain our local stores, it was generally understood that you needed to buy at least the essentials locally. My family loyally followed that motto. However, it was a difficult commitment for many people to keep, and unfortunately it was the lack of community loyalty that would eventually cause our stores to struggle and finally disappear. Management issues also had a serious role to play, as service issues and personality conflicts eventually assisted with the demise of our once busy stores. The temptation for small town gossip can unfortunately disrupt local friendships, and even affect the business climate.

From the beginning, Tyner, like most other small towns in the area, was not blessed with the capacity to serve all the many business needs of the surrounding area. As a result, shoppers would go to other towns to satisfy their additional requirements. Folks would go to larger centres to purchase farm machinery, get major repairs done, or take the family to the doctor, dentist or hospital. For example, there was no café in Tyner, so people went eight miles over to Lacadena or twenty-five miles to Eston or Kyle to have a meal out. There was no "Beer Parlour" as we called them in Tyner, so you went fifteen or sixteen miles south to White Bear or about the same distance north to Plato, maybe even twenty-five or thirty miles to Kyle or Eston for a beer at the pub. As populations declined, our towns joined together to form sports teams for hockey and ball. Eventually, the boundary lines between our rural towns became blurred or non-existant as towns shared resources, including people, in order to survive.

Trips to the city were very rare in those days. However, our family really enjoyed our annual travel day to Swift Current for the "Frontier Days and Rodeo." It was there that we first observed the majestic Royal Canadian Mounted Police Musical Ride. For our family the first trip we made to the city of Saskatoon that I remember was for my brother Eldon to visit an eye specialist. For me, the highlight of that trip was staying in the old Senator Hotel with a chance to go up and down in an elevator, just by the push of a button. Perhaps even a more intriguing experience for me was walking or running for the very first time on carpeted floors. I was accustomed to my mom's shiny-waxed linoleum floors that had turned slightly yellow due to excessive waxing and polishing. I guess that was the very first time I drew any precise distinction between farm and city life.

Our School

The original country schools, scattered nearer to the surrounding farms, like ours at Gravelridge, were closed and amalgamated into what they called "Larger School Districts." Tyner was one of those. That consolidation ritual would repeat itself about nine years later, when it was time to go to Eston for grade nine at the high school.

Perhaps I digress, but I feel obligated to demonstrate how crucial the schools have been to the development and improvement of farm communities. It would not be unreasonable to wonder why homesteaders in a new, sparsely-populated and barren land, being mostly lonely bachelors without wives or children, would ever agree to pay school taxes on their farmland for the establishment of schools. My hypothesis is that those bachelors combined both a community service and self-interest motivation. Their creative and dynamic prairie vision for the future, would solve both the community's academic requirements and the physiological desires of the young, lonely, bachelor farmers. Now lets see how that played out.

Not necessarily with any urgent love for music, a wise bachelor farmer would purchase a piano that he had no idea how to play. It would be shipped out from the city on the reliable weekly freight train. When his piano was safely located in his farm home, as primitive as his place of residence might be, he would be prepared to take the next step. With anticipation he would wait until a certain Friday, especially if he saw storm clouds in the sky. Then, with a shy but determined swagger he would stop by the school about the time of first recess. He did this because he and everyone else knew that the rural school depended on the valuable musical ability of a young female teacher. She would have been hired partly on her ability to make the school come alive with joyous music. With that bit of research in mind, the young, adventurous bachelor farmer would politely and tactfully ask the teacher if she would like to play his new piano. Better yet, he would wonder aloud if she would teach him how to play. In order to do that, it would be necessary for the young teacher to visit the young farmer at his farm home. To make a long story short, almost without hesitation the young teacher, with physical desires of her own beyond her love for music, would agree to a "proper visit" to the bachelor's farm.

By luck and the grace of God, the unpredictable Saskatchewan weather would brew up a real country storm. Hence the young perky teacher would be forced to stay with the also young, hormone-driven farmer all weekend. Needless to say, within a short while the local school board was on the lookout for a new young female teacher who could sing and play the piano. Yet again, a happy teacher had left the profession that produces all other professions, in order to marry an eternally optimistic farmer, with a common desire to raise a happy family—usually with not enough thought about harder working conditions and more irregular paycheques. Hence in the very early days, the rural population began to grow and thrive, in large part because of this important but often overlooked procreation role that schools played. Farmers therefore eagerly agreed to pay school taxes. Come to think of it, my dad, a young bachelor farmer, married my mom who was the young teacher in Tyner.

From our first home in Tyner, a miniature, yellow, cedar-sided, two-bedroom house with red cedar shingles, it was only one block north across the grid road to our friendly, white stucco, flat roofed two-room school. When I started, we had two teachers teaching grades one to twelve, dividing the school into two classrooms, with each teacher handling six grades. The teacher in the high room would be the principal. In those days, our entire school resource library had only one set of maroon coloured, hard covered Compton Encyclopedia. There was no doubt that the higher grades needed to move on to a larger school with more sophisticated resources. There were no on-line resources to access and correspondence classes were not the best option. Therefore, the higher grades began to move away to larger schools like Eston, resulting in the removal of grades at our school.

Tyner School

Nonetheless, I was proud but somewhat apprehensive as I began my journey into education in a small but friendly and caring school. I was happy to begin meeting my teachers, who I usually really liked, and I felt they liked me. I instinctively knew they wanted me to reach my potential. Therefore, I was proud to take my turn on extra duties such as cleaning the blackboard (which to my astonishment was green not black), cleaning the brushes, and even putting up the Union Jack flag. I tell my city friends; "I was so excited when I passed my grade three exams that I couldn't even shave that morning." I think my interest in hand held school bells, which I now collect, began at that time. As students, we loved the fact that when that bell was left in the hand of our likeable school principal, he could increase the time for recess or noon hour, based on our good behaviour and his frequent goodwill. He produced a special musical sound from the way he flipped that brass bell. Like most kids I preferred the ringing of that bell when it signified the end of the school day and set us free.

In my mind I clearly recall how we would golf our way from place to place, even to school and back. Throughout our town from location to location, we had built our private "Salmon Can Golf Course." We dug open salmon cans into the ground so in the summer we went everywhere with a golf club and ball. After a pit stop at home to change our clothes, check in with our parents, and have a snack, there was always lots for us to do.

Our streetlights had a special historical significance for us kids from Tyner. Whether we were at the rink, at the ball diamond behind the school, building a fort in the wood pile at the adjacent farmyard, rafting in a water filled ditch, snaring gophers at our "gopher town" next to the railway tracks, playing "kick the can" on main street, or just riding our bikes, when those lights came on automatically at dark they were the illuminating signal for us to head home, within twenty minutes or so.

Our school would usually have its graduation ceremony at a picnic celebration down by the South Saskatchewan River about nine miles southwest of town. After our school was reduced down to just eight grades, the graduation became a farewell for the students moving on to grade nine at the high school in Eston. The standard parting gift was a nice ring with the black and gold school emblem on it. I remember wearing my ring for about ten years before one day, when I was hauling grain as a young farm hand, I jumped off the grain box to the ground. My ring caught on a screw on the box and yanked it off my finger. Lucky for me by that time it had worn so thin that it ripped off, without taking my finger with it. That painful experience has convinced me as a farmer to never wear a ring, not even my wedding ring.

The school picnic at the river reminded me of what a diverse farm environment we have, contrary to what some people think is just boring, flat, dry prairie. Our graduation celebration took place at what we called "Larson's Grove" in a large grove of maple trees, near a channel branching off from the main South Saskatchewan River. There is a table-like prairie grass area right next to the grove of trees, which is a great area for ball games, sack races, three legged races, a band stand and even a political stump for local politicians or snake oil salesmen to deliver their spin. At the graduation, Jack Rowley's popular orchestra and marching band made up of local musicians would make the grove area come to life with music. The speeches and ceremony part of the graduation event were kept short, as the chairman of the school board in an uncharacteristic and uncomfortable but adequate speech presented the graduation certificates and rings.

Eston HIgh School *(photo courtesy of Cory and Lynne McEvoy)*

Following that abbreviated but respectful ceremony, the real fun would begin. Some people would slip off to the channel for a swim, but be back to pick teams for a ball game. There was always a symbolic ritual to get the teams selected. First, the two biggest and most obvious ball players would be given the high authority of selecting players, one turn at a time, as the captains for each team. You knew you were being judged as a ball player by how much earlier or later you were selected compared to someone else. I made the long journey or evolution from last-picked to captain over the years. Those not into ball would start the sack races and other games which happened throughout the hot, humid, summer day.

About three hundred yards from the grove of maples and the ball field, we were blessed with a very large river-hill, which had a dangerously steep slope on the front. You could not safely drive a vehicle or ride a horse up that slope. However, you could carefully scale your way up that slippery slope with some dexterity and courage, knowing that many participants would fall hard to the bottom. The back of this "killer hill" was not as difficult to climb, so some brave souls went up the back and came running to the front for a circus-like, roller-coaster thrill, sliding down the steep front side. Many, many, many times, this hilarious and exciting event ended in a broken collarbone, a lost tooth, a broken arm, a sprained ankle, a sore back, or something less severe but just as memorable. Usually the doctor in Eston knew first hand what day we had our picnic.

Finally, the day would end with a wiener roast supper by a large controlled bonfire, complete with potato salads, jello salads, vegetable trays and fruit trays, as well as cakes and pies. Drinks of "Kool-aid" or "Freshie" as we called it, were mixed from a thousand gallon tank of cold spring water, delivered on a local farmer's water truck. Then there was the prize of the day, which was also the last activity before everyone headed for home. The final prize of the day came when the local storekeeper unloaded from his truck a large tan-coloured, heavy canvas, cylinder-shaped container, designed to insulate its contents from the heat of the day. Inside that container he would methodically scoop out more-than-generous mounds of hard ice cream onto man-sized cones. What a friendly way it was to end a wonderful day.

The only thing that could make it better and quite often it happened, was when everyone was safely home, the dark rain clouds, with bright chain lightning bolts and ear-piercing thunder, would deliver a much needed June rain and cooler temperatures for the crops, while we listened from the safety of our beds. Folks would call it a "million dollar rain."

Our Rinks

The first skating rinks of the winter season for us kids before the indoor rink was flooded, frozen and ready to go, were two dugouts located one at each end of our town. As soon as they froze over in late fall and were considered safe to be on, we would be out there with a shovel or two, our skates, our hockey sticks and pucks to clear the snow and get off to an early skating season. We would sit on an old pail we brought with us to put our skates on and at least two of us would offer up our overshoes for goal posts. In those days there were only six teams in the NHL and two of them were the Montreal Canadiens and the Toronto Maple Leafs, both from Canada. Then there were the Detroit Red Wings, Chicago Black Hawks, New York Rangers, and Boston Bruins. Those were the days when NHL Hockey really was Canadian and the Montreal or Toronto teams often won the prestigious 'Stanley Cup." Usually my friends had either a red Montreal Canadiens or

Tyner rinks

a blue Toronto Maple Leafs sweater to wear. Therefore when we chose up sides for a game of "shinny" we would all throw our sweaters in a pile at centre ice to be divided-up and shared between a red and a blue team. Who would be Rocket Richard, Jean Beliveau, Boom Boom Geoffrion, Plante or Harvey from Montreal and who would be Johnnie Bower, Dave Keon, Tim Horton, Kelly or Armstrong from Toronto? I loved Jean Beliveau, so I always had a large number four taped with thick white tape on my red Montreal Canadiens jersey. "Let the cold windy game begin."

Just south across the hard dirt street, next to our first little Tyner house, were the two wintertime recreational loves of my early life. Those were our curling and skating rinks. Our long, single-sheet curling rink was attached to a very inviting and cozy, wood and coal heated waiting room and kitchen. Curling in rural Saskatchewan at that time was a way of life, perhaps even more than today. In our town on that one-sheet rink, there were three draws or games scheduled every day of the week, except Saturday and Sunday. One game at four-thirty just after school, then one at six-thirty just after supper, and the last one at eight-thirty ending at ten-thirty at night. Our local curling club was very welcoming to young people like me, and eager substitute curlers were always needed. Some days, I recall when I was somewhat older, I got to play in all three draws, either on my regular team or as a substitute with another team.

After a couple of years in town my family moved two blocks to the west on the same street into a bigger home in a larger yard complete with nice trees and a large garden. In the broader scheme of things we were just two short blocks further from the rink and the same distance to the hall and stores. Our new, white, wood-sided home with green trim had a form of running water to the sinks and tub which was a first for our family. However, it didn't have water and septic tank for the toilet so it had to be carried out and dumped. Not a chore anyone liked but a job that needed to be done. Living without a flush toilet might seem impossible but we did it.

Those were the days of hip-length, heavy-knit, matching curling sweaters, and super wide corn brooms going "swish, swish." down the ice as curlers let the forty-pound granite curling rocks go from the hack. That "hack," as they called it, was a rubber foot holder, frozen into the ice to give a solid send

My second Tyner home

off to the rock. Later, the more modern, narrow broom that went "smack, smack," came into vogue, long before the quieter push brooms of today. The younger curlers adopted the slide, which effectively shortened the distance to the far rings. You could slide as far as the hog line, which was about twenty feet out from the hack, before letting your rock go on its own, heading toward the coloured rings called "the house" at the other end of the rink.

During the first week of February every year, curlers from all the other towns within a thirty-mile radius or more would come to Tyner for our annual curling bonspiel. Maybe your team would win a trophy, a set of flashlights, a case of oil, or curling brooms. During these prairie events, you could not beat the home cooked meals provided by the communities. For supper each evening, there would be a delicious menu featuring one of roast beef, ham, turkey, chinese food, etc. At noon each day there were homemade soups to die for. During our famous Tyner Bonspiel one time, during a long day at the rink, I ate one dollar's worth of homemade pumpkin pie, with a ski-hill of real whipped cream on top. The cost was ten cents per giant piece.

Yes sir, that was ten pieces of pure nutritional pleasure! I still love pumpkin pie and whipped cream today, even at about five dollars per stingy slice.

Right beside the curling rink, about forty-five feet away, but not attached to it, we had a lively skating rink. Going to it, you carefully glided on an ice trail, after putting your skates on in the toasty, warm, waiting-room section of the curling rink. I vividly recall the large, iron, pot-belly stove in that waiting –room, which would get red hot from the flaming wood and coal taken from the coal bin in the attached shed. Usually, we had a self-serve strategy, where we all took turns building and refuelling that somewhat primitive but effective heating system. There was a custom made metal water tank, which fit like a coat around that hot stove, to heat water for flooding the curling ice.

I remember skating or practicing hockey for hours and hours and hours on Saturdays and Sundays as well as other weekday after-school hours. As boys, our negotiation skills had an opportunity to flourish as we skilfully negotiated with the girls as to when we could play hockey with sticks and pads, or when the girls could skate without the fear of getting hit by an out of control puck. That was often a challenge in diplomacy.

In my high school days, I vividly recall doing my part assisting my dad and three or four other volunteers to flood and make ice for both the curling and skating rinks. Trucks and tanks would hall water to the storage cisterns, where we would hook up an electric pump with many long hoses joined together. After several floods, the ice surface would be thick enough. Then we would draw and paint the circles or lines on the new ice. The curling ice needed more attention, as it needed to be scraped and flooded with hot water to make it harder and smoother. This process of making natural ice could not start until well into November due to weather, which meant skating and curling could not begin much before Christmas. Then by March, the ice disappeared. It was a short but busy season, which meant our rinks were very well used as a major part of our farm culture.

My dad and his team, including myself, were true volunteers who, without pay, after our own work was done, came together to help our community. That simple definition of volunteer became more meaningful to me later in my life, while living in the city. I was confused and shocked to learn that often city-style volunteers are just regular paid employees who are lent by their employer to a so-called "community volunteer project." Then I was more disgruntled to learn how those city folks were given and actually accepted government medals for this type of phoney volunteer contribution.

Since by now you know me quite well, there is something I can now tell you, if you promise not to tell anyone else. OK. There is a deep, dark, dirty little secret, nurtured and carefully concealed by bone fide former Tyner residents, about the physiology of our beloved old skating rink. Like a defacing affliction, it was not something we were very proud of, but we came together to reluctantly live with our collective disability. I hope that offering you this painful revelation will not expel me from my heritage. The whole truth and nothing but the truth is this: Like a proud erection gone soft, our skating rink started to sink in the middle of the roof with the high probability that it would soon come down. It would no longer be able to safely perform the most basic acts of procreation necessary to sire new hockey players and figure skaters. For safety reasons and to save our manly pride and joy, our embarrassing community rink "erectile dysfunction" needed to be cured. Quickly, almost in the blink of an eye, a typical "barn raising bee" was organized in the tradition of farmer helping farmer. Soon, our droopy rink was repaired by a competent local platoon of legitimate volunteer carpenters and helpers, using large construction machinery and piles of heavy building materials. However, from that time on until both the rinks were torn down, there was a large square, laminated wood pole dead centre in the middle of our rink. That giant one foot square post, for reasons of safety, held up a giant truss rafter assembly that safely kept the roof up in place. That post would serve as an impartial, solid third centre man, equally fair for both teams during hockey games. As you can understand, over the years as a town we fielded many sarcastic but playful criticisms about our unique rink design. So now you know the rest of the story.

Our Grain Elevators

Somebody once stated that the prairie grain elevators located about eight miles apart in every little town were built to serve as protection from the harsh prairie wind. Each elevator would have the name of the town printed on it for the world to see. In the days of carefree parties and night travel from town to

The Saskatchewan Wheat Pool elevator
(photo courtesy of Wayne Corbett)

town you could wake up in the morning, look out at the elevator and discover where you were. Tyner had three traditional grain elevators, which defined its south profile. There was a dull maroon coloured Sask. Wheat Pool elevator, a faded grey-white Federal Grain elevator, and a bright orange Pioneer Grain elevator. They all had a long reach into the sky, about eighty feet high.

In order to be a legitimate farm boy, you needed to be able to describe without hesitation how a grain elevator works. Let me give it a try. Any grain elevator, no matter which company owned it or built it, is operated in much the same way. The grain is delivered by the farmer's truck into the elevator through the driveway entrance, where the grain box is weighed by the balance scale, then lifted by the hoist to dump the load through a steel grate on the floor into the pit below. That pit is sloped to slide the grain over to the boot, as it was called, at the base of the leg. That leg is the long narrow enclosed shaft, complete with grain cups on a wide rubber belt, powered by the diesel engine to elevate the grain up to the top of the elevator. Once up to that height, gravity takes hold to drop the grain through a selection of pipes directed into various overhead bins, later to be further dropped through another pipe into the grain car spotted under the loading spout on the railway track. Sometimes this elevation process drops the grain into an attached storage facility called an annex. That annex may be of wood frame construction or a steel bin. When it comes time to ship grain from the storage annexes, an auger at the base of the annex moves that grain over to the boot as we just described at the base of the leg, where it goes through a repeated elevation process where it also goes into the grain car on the railway track. More modern or upgraded grain elevators were adapted to be powered by electricity in place of the original diesel engine.

Friday was usually train day, when the old box cars or later the newer hopper cars were spotted on the side rail nearest the elevators, ready to be loaded with grain. They would be shipped west to Prince Rupert at the Pacific Coast, east to Thunder Bay on the Great Lakes or north to Churchill on Hudson's Bay. From those terminals, it would be loaded on ships for world markets. The Canadian Wheat Board had been established as the farmer's window on the grain trade as an active player in the global market place. The grain elevator companies would be paid for handling and storage for CWB sales.

A regular source of income for me as a boy was helping the elevator operators or "elevator men," as we called them, in the busy times. Some of my jobs would be lifting the pails to help clean out the "boot" at the bottom of the leg which I described, or helping "cooper the box cars" which meant putting the doors in place, or shovelling grain taken out of the "annex" to be loaded in the rail cars. As I got older, the level of assistance I was able to provide got more involved and eventually included helping with the calculations and reporting procedures on "books day," which also came on Fridays. Those earnings paid for essentials like my ball glove, hockey equipment, movie tickets, some clothes and other treats or necessities including practical gifts for my parents and my siblings. I remember how I would chip in with some of my earnings if the family wanted a new piece of furniture, or when we got my dad a better set of golf clubs.

Our Tyner Memorial Hall—An Integrated Facility

The hall was built after WWII as a revamped grain storage annex, out of respect for the heroic soldiers from the district who returned from war, or sadly, gave all they had for our country including their lives. Perhaps what I remember most about our hall was a large sacred collection of pictures prominently located on the north wall of the main open area. It was comprised of individual portraits of local service men and women, featured in their dignified uniforms from the Army, Air Force or Navy. My dad had his place in that glorious display, dressed in his army garb. He had volunteered to serve, and with his farm expertise was selected to drive an ammunition truck on the front lines during his four years of service. It was quite true, as he often joked, how he supplemented his grade eight education in the school of hard knocks, serving our country in Sicily, Italy, France, Holland and England. As I recall, he didn't want to talk much about his service and he always said he was not a hero. He would say it was a job with a paycheque, a way to see the world, and his country needed him. Perhaps the most penetrating message he offered me about his time at war was his frequent comment: "War is bad business." To me, he always was a hero and everyone in that picture deserved all the respect and praise they received.

Our hall was the church and Sunday school, a movie theatre, a live theatre for Christmas concerts or plays, a dance hall, a badminton court and a general meeting place. It was the preferred gathering venue for Saskatchewan Wheat Pool and Canadian Wheat Board suppers and meetings, Boy Scout and Cub meetings, Tyro boys club and Canadian Girls in Training (CGIT) meetings, Ladies Aid, Homemakers or United Church Women's meetings, along with rink board and church board meetings. It was clearly the nerve centre of our active community. Truly, it was an integrated facility.

Let me try to remember some of the details that describe our old town hall. On the exterior of our hall, the roof was covered with red painted cedar

Tyner Memorial Hall

shingles later replaced with red asphalt. The siding is a light grey masonry type. I recall as you walked over the wooden platform and up the stairs into the porch, you came to a set of brightly painted white double doors. Going through those doors, you would see to your left side a sliding window which served as a ticket-taking kiosk, before you pushed open the shiny, swinging, varnished interior doors, entering the hall proper. Your first impression would be "Man, this place is so fresh and clean, from the polished golden hardwood floors, up the clear varnished wood-grain walls to waist high, then with upper walls of soft textured beige wallboard. It is so high and open like a cathedral, with several large, elegant, hanging, bell-shaped ceiling lights." Off to your right near the entry area is the girls'/ladies' cloak room, and on your left, the boys'/men's cloak room. Above the boys'/men's room and up the narrow stairs is the cramped movie projector space. Next to the girl's room is the large attached kitchen including fridge, stove, sink, and giant coffee maker. The conventional cupboards were accented with special narrow shelves custom-designed to hold loaded cake and pie plates. Stacked on the cupboard you can find several red and grey metal trays, with pictures of the Queen, Santa Claus or flowers on them, ready to serve the generous and delicious lunches associated with rural community life.

Moving to the front of the large interior, there is a famous players stage, complete with a sophisticated red velvet stage curtain, which is opened or closed by a complex assembly of ropes or cords controlled by a wheel with a turning knob. Behind and beside the stage, there is a somewhat cramped but adequate behind-the-scenes space, accessed by stage doors needed for theatre productions. Up on the back wall of the stage itself hangs a very professional roll out movie screen, ready to meet the curious eyes from the audience, sitting on golden varnished plywood and metal chairs. Next to the stage, there is a stairwell down to a small basement, where the all-important bathrooms are located. Just in front of the stage on each side of the room, there are fire escape doors with dark red exit lights above them. You will notice two-inch-wide black lines painted on the main floor area marking a badminton court. The sharp black lines stand out against the golden-oak-varnished hardwood floors. Then, consistent with the dominant golden wood decor, the simple furniture consists of many smooth, varnished wooden stacking chairs and matching tables, which were also made for stacking or joining together for larger areas.

I must not forget to mention the annual hall clean-up day. That was the day when many ambitious and committed volunteers came together to scrub and repair the hall, including hauling water for the cistern and polishing the light grain hardwood floors, as well as doing any other touch-up jobs seen as necessary. For me, in my early days, the lunch was what attracted me to help with the work, along with my parental directions. Later, as I matured toward manhood, I think it was the low-cut cleavage revealing tops and stimulating miniature short-shorts worn boldly by the ladies that motivated my participation.

My very first memory of the Tyner Hall is when, as a family, we would go into town from the farm to see a movie. Train night was when the movies would regularly arrive for Lee, our loyal projectionist to prepare to play Saturday night at the hall/ movie theatre. These were great occasions. Keep in mind that in those days we didn't have television. I can still taste and smell the popcorn along with that great big bottle of orange crush. The giant silver screen would be rolled down at the back of the stage, and when you looked way up, you could see the flying dust coming off the projected light shooting out from the little square hole in the projection room all the way onto the screen. Before the movie, a young Fred Davis who later became well known as a CBC-TV host on "Front Page Challenge," would host an interesting "National Film Board Documentary," based on news of the world to start our evening's entertainment. Following that, there would be a short intermission. We would soon be beckoned back to our seats by cheery music introducing Woody Woodpecker, Mickey Mouse, or Bugs Bunny for the pre-movie cartoons. Popcorn and pop in hand, we settled in for the movie.

I remember as a young lad how we came prepared to meet Roy Rogers on Trigger and Dale Evens on Buttermilk, with Bullet their trusty German shepherd dog and bodyguard. Just like it was yesterday, I can still see Roy with Dale at his side, driving the stagecoach, transporting a shipment of gold for the bank. To Roy's and my total surprise, along came two mean looking masked men, ready to rob them of the gold. One big guy with a six-gun revolver aimed at Roy said: "Give us the gold or I will shoot." Now, Roy was a real cool dude who could play the guitar and sing a good country song. He calmly says to these two bad guys: "Would you fellas like to hear my new song first?" Just before they could reply, he grabs his tan-coloured acoustic guitar. Without hesitation or visible fear, he points the upper frets of his guitar at the robber and with a loud bang shoots the gun out of the robber's hand. Crafty old Roy had a hidden rifle inside the guitar. Then Roy, without even moving his elegant white hat, drew one of his own six –guns with the pearl handle, and pointed it to keep the robbers in tow, while he and Dale tied their hands. Bullet stood by just to make sure those robbers with the black hats didn't try anything stupid. I was so impressed that for weeks I wanted a pair of brown corduroy jeans just like my hero Roy Rogers wore in that movie. I never got the jeans but I did get a Roy Rogers lunch kit to take to school. Some of the other cowboy heroes I had from those movies were The Lone Ranger and Tonto, Gene Autry, Hopalong Cassidy and Rex Allen. Snow White and the Seven Dwarfs, along with Tarzan, were also non-cowboy favourites.

Our town hall was prepared when necessary to accommodate christenings, weddings, or funerals. For the most part, it served as a United Church of Canada place of worship, affiliated with nearby United Churches in White Bear and Lacadena. Lacadena was where the manse, a home where the minister lived, was located between Tyner and White Bear. On some less frequent occasions, a small group of Seventh Day Adventist worshippers would also use the hall for their church services on Saturdays.

Tyner, as I knew it, was generally a clean-living, churchgoing, quiet rural community. My family was active in the United Church, where I earned thirteen attendance pins, received each year if you attended a set number of services. I recall taking my regular turn with my friend Darrell, who was about my age, passing the collection plate up and down the rows. Nobody that I noticed ever took anything out of the plate, and most people contributed.

My mom taught Sunday school, sang in the choir and was president of the UCW (United Church Women). Her church quartet of singers even went to Swift Current, about eighty miles away, to sing on CJFB television. CJFB was a CBC-affiliated station. My dad was secretary/administrator for the Sunday school and handled the finances or books along with serving on the church board.

Our church in Tyner had an informal but rigid routine where the same cars parked in the same spots outside and folks sat in the same seats inside. There was no officially assigned seating, but we all knew who sat where. One time, I recall that a friend and I caused much concern and embarrassment for our parents and our extended church family when the preacher abruptly told us to move apart. We were making too much noise by talking. It is difficult to be perfect when you are young, so I guess we didn't try very hard sometimes. Even with my imperfections, I did enjoy the old gospel songs and several of the Bible stories. However, to this day I have never understood how religions can cause so much grief and hurt in the world, including death for their foes.

For my take on religion I can truly state that the main problem with Christianity is usually the Christians. Put another way, I sometimes say: "The bigger the book—the bigger the crook." Most of the doctrines have merit and even match my views on life and common humanity, but personal interpretation guided by individual privilege, greed, and hypocrisy often convolute the purity of any religion. For me, it is more about spirituality and acts of kindness and sharing than the organized ritual of dogmatic self-serving religions.

I like to tell this "nearly true story" about one of our preachers. As I mentioned before, almost everyone in our town went to church on Sunday. One time we had a preacher who rode his bike everywhere he went. However, this one day my dad and I are walking down the street when we meet up with the preacher, who is walking. My dad asks him where his bike is. He replies by saying his bike was stolen and he didn't know what to do. That was when my dad advised that he should preach a service on the Ten Commandments. When you get to Commandment number eight about "Thou shalt not steal," you should just yell it out at the top of your lungs. Then closely look at everyone's face and you will find the guilty one. The preacher agreed to do that sermon next Sunday. As we all sat in the pews, we heard the preacher give his sermon. However when he came to commandment number eight, he just read it quietly and carried on. He certainly didn't yell it out. After the service, my dad asked the preacher why he didn't follow his free advice. The preacher said he would have, but when he got to Commandment number six about "Thou shalt not commit adultery," he remembered where he left his bike.

In rural Saskatchewan, including Tyner, historically there has been no discomfort with holding Christmas concerts as part of the school program. To my knowledge, there has been no great objection to that Christian tradition from non-Christian organizations or individuals. It has been seen as a great opportunity to implement learning related to stage productions all the way from writing, drawing, preparing, practicing, acting, playing music, and building the props needed for an entertaining performance.

At our school, it was usual procedure near Christmas-time for the students to begin attending school, not at the school, but in Tyner Hall. This change of school to hall venue would happen about December 1 or so, with a view toward holding the annual Christmas Concert around December 21, which would be the last day of school before Christmas Holidays. We would have about three weeks of intensive learning in what we called the performing arts. Parents, teachers, and students enjoyed this part of the school program and found it very productive, with educational value.

My most vivid memory of this school-joins-the-hall occasion was when I was in grade eight. When we went to the hall to prepare for the Christmas concert, it was like everyone was in the same grade with two teachers for three weeks. While the focus was on the concert, tables were set up for regular studies so the regular courses could be advanced when we were not part of the concert practice. Our principal was amiable and fun-loving, but he also meant business and wanted us to succeed by getting our school work done. The other teacher also was firm but very good at developing the art and music for our concert, so in many ways she was the boss. She could really make that innocent but dominant piano shake the floor with enthusiasm. Together, our teachers made a very good team, and they always had our best interests at heart.

That particular year, I remember the TV show called "Bonanza" was on the air. It was a western, with the dad Ben Cartwright and his three grown cowboy sons Adam, Hoss and Little Joe, who all lived on the Ponderosa Ranch. Hop Sing, a likeable Chinese cook and housekeeper, provided the domestic services. Adam, the eldest son, was a slim and handsome ladies' man who could sing a lilting love song and serve as a dedicated crime fighter when needed. Hoss was a portly, happy-go-lucky and friendly cowboy who was instantly recognized for the big white round-topped cowboy hat he wore. Little Joe was the young, cute, heart-throb who really could attract the ladies and all the problems that came with them. Ben, the father, was clearly the "father knows best" type as ruler of the roost, with a large dose of wisdom and compassion.

This popular contemporary television show would be the inspiration for a stage play I would have the starring role in for our Christmas concert. Given my abundant stature and jovial personality, I was typecast to play the part of Hoss Cartwright. I still don't know whether my selection was a complement or not. Anyway, it was a challenge because as a cast, we also had to write the production, produce it and act it out. It was a great learning experience for me and everyone else. Even with our show biz limitations, we did the best we could. We learned a lot, we enjoyed the experience, our parents and friends applauded, and Santa Claus never forgot about us. Tomorrow the Christmas holidays would begin.

I believe that small town dances did more to define the rural identity and culture than any other community event. If the walls of the Tyner

Memorial Hall could speak, they would have hundreds of tales to tell about local dances. A live band of strings, brass, and drums, often filled the overheated and sometimes stuffy interior air with stimulating music, while exuberant couples clinging together in each others' arms meticulously and orderly spun each other around the hall, at a variety of rythmic speeds well into the early morning hours. It was an experience you should not miss.

Stories go back to the days when folks came by horse and sleigh or buggy to all night dances before they rested a bit and returned to their farms. Either as a participant or spectator, the joy of the event was tremendous. Slow waltzes, fast polkas, outrageous butterflies, clever jive and other variations of those dances were executed within a variety of levels of perfection and complexity. Some folks quite sheepishly will admit that the level of intensity and innovation beyond the regular steps was directly related to the degree of alcohol intake. I think I can lend my personal experience to that declaration.

The physical arrangement in the dance hall was set up with chairs around the edges of the hall next to the walls and closer to the front or stage where the band was set up. That was where the ladies sat with their purses and bags secure but openly sitting under their chair. By choice, the men generally stood at the back of the hall in what was affectionately called the "stag line." It was close to the door.

From all reports and some personal experience, I can tell you that it takes a sufficient dose of courage to advance from the stag line up to the chairs near the front of the hall where the band is located, to ask a particularly attractive and perky lady for a dance—especially if your dancing skills left quite a little bit to be desired. Those with partners like wives or a date would not have to live through this temporary but uncomfortable stress. They would be doing their "duty dances" and even if they didn't harbour a great deal of enthusiasm for dancing, they dared not show it. I guess this must be where the old political comment stating: "You need to dance with the one who brought you" originated from.

Something not well known is how important the scheduled breaks were that the band would take between the sets of dance tunes they played. People needed time to communicate and refuel, if you get what I mean.

Historically, up until recent years there was no booze allowed in the dance hall. For many, it was necessary to periodically move outside to the parking lot, where certain vehicles quickly turned into booze vendors, or in this environment, courage builders. Usually this self-medication would need to be repeated a few times before the desired results were evident. That was when the fun started. When the music resumed inside, it was time to move back into the active and joyous dance hall. It was time to shake a leg or do the soft shoe. You may be amazed at what a little artificial bravery can do for a hormone-driven young man.

By now that quite ordinary girl on her chair near the band had acquired irresistible beauty. It was time to make your move. With all the swagger and poise you could muster, you would emerge from the safety of the stag line, and slide with the frantic beat of the music up to the lady of your dreams. "Would you like to dance?" was a common pitch, applied with a newly-found twinkle in your eye and a genuine smile of anticipation on your face. Sometimes that alcohol-sponsored gesture did not work as well as expected, and the lady would say "No thanks." Then it was time to think quickly on your feet. Should I ask the girl beside her? Generally that was not a suitable remedy because when the girl you first asked said no, usually the gals beside her would refuse as well for some unknown reason. Sad as it might be, there was no alternative but to complete that long, long, long, humiliating walk past all those critical observers, back to the stag line and then sneak out for some more liquid painkiller. Casey had struck out.

Sometimes you would have more success. Not always was the result of your journey to the front with your liquor-induced invitation so devastating. Many times that adventurous and also very human young lady would say yes to that first dance. You might be on your way to a pleasant and rewarding relationship. At the very least, you joined the excitement and companionship of the night, and together you enjoyed the rest of the evening, with a delicious lunch and coffee. Then their was the much anticipated cheek to cheek, slow and easy last waltz, which prepared you for the journey home or maybe off to an after dance party. All of this made your previous stress all worthwhile. For added excitement, that trip to a nearby farm for the after-dance party might be in my 1965 white two-door hard-top Chevrolet Impala, with hot red interior, along the railway track to avoid the strong arm of the law.

Our once-active Tyner Memorial Hall has been gently and lovingly moved and restored over at the Eston Riverside Park, about twenty miles away to the southwest, on the edge of the beautiful South Saskatchewan River. Much to my pleasure and to the pleasure of many others from our community, the move allows our Hall to continue to serve as a multi-purpose integrated facility to this day in its lovely new location. However, as the last remaining community-owned structure, from the happy life of my hometown, went down the road to its new home, it was sad to see it disappear. Now, it would be only our memories that would keep our friendly town alive in our hearts and minds.

In rural life, not unlike life anywhere else I suppose, I know it can hurt when someone or something moves away. As a boy, I remember how difficult it was to accept the loss of a good friend when his family moved to a new job and new location. Even with good intentions to keep in touch, I recall it usually didn't happen to the extent of making up for the loss. For some reason, the removal of our Tyner Hall, truly a real integrated facility and jewel of our town, reminded me of the time when a neighbour girl I really liked moved away from our town with her family.

With hindsight and reflection, matured by the passage of time, I can now admit that I had a rather clumsy but severe crush on this stunningly attractive and accomplished neighbour girl. She was a couple of years older than me and lived right across the alley from our place. Unfortunately, she never did take me very seriously beyond her naturally polite nature, although I wished and hoped otherwise. Even though I could not fully understand her at the time, she clearly was interested in older guys with a car and more freedom.

In fairness to my heart and growing passion, she was completely worthy of my youthful admiration. She was very good at school, and very proficient at sports, including track and field, with running and the high jump being her best events. In addition, she mastered senior levels of conservatory piano education with many hours of practice, often within hearing distance for me as I loitered out in our yard. I would speculate that she played her best when she thought I would be listening. My mind would scramble at times to invent reasons or excuses for me to visit her place or meet up with her somewhere.

Compounding my confusion, this girl had a very beautiful, slim, lanky body complemented with a lovely smile and catchy personality. All of those attributes easily caught the attention of a bewildered lad in the throes of puberty. Thinking back, I now believe her parents had me figured out but never outwardly objected. One day, this beautiful neighbour and her parents, who were good friends of my parents, let us know that for her dad's employment they would be leaving Tyner. The loss of that family was difficult for our family and our town, but for me it seemed like the tragic loss of life's true potential.

Now the school is gone, the rinks are gone, the stores are gone, the post office is gone, the elevators and railroad track are gone, and our Tyner Hall—the nerve centre of our community is gone. Over time we were losing my hometown and much of our rural way of life, one person, one family, and one building at a time. For some strange and morbid reason, as homes, businesses, or service buildings like our hall disappeared forever, farmers replaced them with large, cold, hard steel, silver grain bins to serve the fewer but much larger family and corporate farms. Those rather unsympathetic and lifeless structures that symbolically have taken over, displacing my past, now serve as drab, unmarked headstones, representing the death of my hometown in the graveyard of time. All that is left for me are my memories—*"Memories of a Suitcase Farmer."*

three

FARM COMMUNITIES HAD FUN TOGETHER

Lately, as I have stopped to think more about my hometown, Tyner, Saskatchewan, it has dawned on me that while I had a comfortable hometown it didn't actually offer us all the services that we desired even back then. The humble truth about most rural towns is that they are always part of a larger constellation of towns, working and playing together to collectively provide as many social or business services as they can. Going back as far as I can remember, the unstoppable depopulation of rural areas has prompted our smaller communities to join together and pool their resources as a matter of survival. For example, the best hockey I ever played on a team that had the most success of any I was ever part of, happened when Tyner, Lacadena and White Bear combined our best players. That merger allowed us to play in a league of larger, more competitive towns. We were the "Legionaires."

The "Chinese Café"

My hometown of Tyner didn't have a cafe of its own, but Lacadena, Plato, White Bear, Eston and Elrose did, so we always adopted one from another town. Probably those cafes were not much different from hundreds of others that dotted the prairie landscape at the time. Any story about rural Saskatchewan would not be complete unless it would reflect on the "Chinese

Café" as an integral part of the rural culture. My memories of those premises have all blended together over the years.

Like almost all other such establishments, the building was usually quite narrow but very long, stretching back from the concrete sidewalk on Main Street in the centre of town. The front of the building below the large windows had a huge red and white Coca Cola sign, ready to greet you as you came to the well-worn screen door that banged hard when its spring pulled it shut after you got inside. Everything in the café, as I recall, was either red or grey-silver. The booths we collapsed into were very comfortable, made of soft and smooth red vinyl, with grey chrome tabletops on strong silver metal legs—all easily cleaned with a wet cloth. The exact same look came from the customary red vinyl, round, swivel stools which lined up along the long and narrow countertop table. Very close to those stools and table there were the cupboards, sales shelves filled with goodies and smokes, the pop machine, the jukebox and the heavy-duty cash register. At the far end of that area there was a set of swinging saloon-style doors leading to the mysterious, usually off-limits kitchen and the stairs leading up to the second story living quarters.

Putting a dime in the jukebox, from a song selection mechanism, right at your private booth was a delight. That was how you got to select the musical background you preferred for the gossip and nonsense, which would usually take place at the various booths or stools. It didn't seem to take much of a disturbance, as I recall, to send the slightly-built dark haired lady waitress/co-owner of the Café, into a wicked tailspin of rage. Patrons would need to be prepared for their frequent but temporary eviction from the premises. Sometimes, I must confess we would go there with a plan to tease our fragile waitress into a fit of near insanity.

Usually "the chinaman," as he was known at the time, had a flair for business and making money that extended beyond just running a legendary rural "Chinese Cafe." Not many people got to know Sam very well, but those who visited one-on-one with him would on occasion gain some very interesting insight into his character and his sideline activities. Sam was a gentle family man with a head for stock market investments, and a compulsive shine for gambling. He was known to find his way to the city and the thoroughbred or standard bred racetrack at least a few times per year. On those same trips, he would converse with a stockbroker to place his bets in the equity market. Nobody really knew the results of those ventures, but he always seemed to be better off then we thought the café could provide. All we knew was that he always drove an almost new, more than average priced vehicle. It dawned on us that his family had an obvious income stream greater than serving chips and gravy, pop, hamburgers, and his famous milk shakes could possibly justify.

Sam was not afraid to drift off the straight and narrow in order to make a buck or two. Sam's eye for entrepreneurship spotted a niche market. Put simply, he became the local under-the-counter bootlegger. Because he had regular local customers who could benefit from his services, he was able to keep that little secret down to a rumour and just harmless suspicion.

I guess there was one close call for Sam when the RCMP sent in a young fellow from Kyle to pass some marked money to Sam for a bottle of illegal whiskey. Nobody understood why Sam sold him the whiskey, because he was not a regular trusted customer, but he did. However, the young guy by mistake did not pay with the marked money, so Sam got away with another sale. That was a close call, but it didn't deter Sam in the least. To my knowledge, Sam never ever got into any trouble for his venture.

Another of Sam's endeavours did not turn out so positively in the name of the law and the RCMP. In those days there was no hunting season for Sandhill Cranes, a long necked, long legged, skinny migratory bird. Sam was well known for his powerful ten-gauge shotgun, which had more range than the usual twelve-gauges that all the other hunters used. Sam loved to shoot Sandhill Cranes even though it was beyond the law. Story has it that one evening just before dark, Sam was sitting in a bull berry bluff on a sand hill down by the river, waiting for the cranes to head back to the river. Sam and his old blunderbuss managed to down six long-neck cranes with four bomb-like shots, into the nearly dark sky.

Unfortunately for Sam and maybe the future "craneburger buyers" at the cafe, the game warden was sitting in a sly hiding place in another clump of bushes not too far from Sam's hunting blind. Needless to say, Sam paid the price with a heavy fine, and his trusty ten-gauge shotgun was confiscated along with the removal of any hunting privileges for at least one year. Maybe Sam's luck with the law had run out?

The Co-op Service Station

Soon after the local Chinese Cafe closed down, somebody got the bright idea to set up a coffee shop in the co-op service station. It was a busy little place to buy gas at the pump, order bulk fuel for the farm or get an oil change or new tires. There were even a few hardware supplies you could pick up. The hoist in the high profile side of the building could lift up your vehicle to check and repair breaks, exhausts, springs or shocks and drive train. The low profile half of the establishment was the hardware area including the sales counter, display shelves, bathrooms and large windows looking out at the gas pumps. The whole building was wrapped in wide cedar siding painted a sharp white colour with a prominent green trim. The roof of ordinary cedar shingles was also a dark green. A staff component to run the operation consisted of Eric the manager, Lyle the bulk fuel truck driver and Ernie the service guy.

The new coffee shop would only be a couple of tables and some hard wooden stacking chairs next to a large coffee maker on the nearby cupboard. Before long, this retreat for the recently homeless coffee crowd started to sparkle with country wisdom. Hours would be devoured by this self-proclaimed mechanical crew, analyzing the function of a single-bolt, belt, or bracket on a common machine.

It was easy for these laissez faire experts on both business and pleasure to keep a close eye on staff at the co-op, especially Eric the manager. As life would have it everybody makes a few mistakes and Eric unfortunately was no exception to that rule. The critical coffee shop crew were soon to comment on any minor slip up that Eric or his crew would make in a busy workday.

Jack, the new president of the co-op board, was more critical than most. He was recently asked to join the board and was immediately elected as president because some folks thought as a larger farmer he would bring his fuel purchases over to the Co-op from Esso. Jack had said himself that he didn't really believe in the co-op principles but he would take the position because a good businessman was needed to shake things up a bit. Almost his first order of business would be to advocate for the replacement of the staff, "to get some people in here who know what the word profit means" he vowed. Jack would proclaim right at the coffee tables: "We need to forget about those hand-shaker co-op principles of democratic control, or extra quality and service. We need to get down to making a profit." He would start that process at next week's meeting by making a motion to fire Eric and Lyle. The majority, like sheep without much thought, supported him and the co-op was quickly launched into a pointless recruitment strategy to attract new "high class" personnel. The first task in Jack's process was to produce a sophisticated job description outlining the ideal candidate and converting it into a job ad. Then the advertisement was placed in several local and provincial papers. Meanwhile, Jack and his buddy Johnny decided to actively run the place, even though they both still made their own farm fuel purchases from Esso the competition.

Three or four potential managers applied for the job and they were all invited to an interview that Jack and Johnny presided over with an iron hand. All of the candidates interviewed had graduated from the "Co-op College" in Saskatoon headed up by Harold Chapman. Harold was a long-time educator on co-op principles. Harold is someone committed to the idea that co-ops can be profitable and still remain true to their principles of local democratic ownership and control, as well as provide superior customer service and good member relations. Nobody knew why the other members of the co-op board, with roots more deeply entrenched in the co-operative movement, stood by and let Jack call the shots. Jack and his sidekick Johnny knew best and made it clear that profit at any cost was their mission.

Jack made it perfectly clear to the candidates that as long as he was in charge, profit was the one and only objective. After the interviews, an offer was made but the candidate rejected it following a visit to check the place out and talking to some of the local people in the community about the board members. Two more offers were made and two more rejections occurred. Things just weren't coming together the way Jack had promised. The membership soon began to withdraw their business from what they saw as a co-op that had lost its way. Within four months the co-op fuel delivery and service centre was closed forever.

The Sports Day

When I was young, I know I really hated it when it rained heavy on a Friday or Saturday in June or July. There are good reasons for anyone to feel this way when you could miss any warm summer day. For me, what was special on those Saturdays was the string of "Sports Days," which always included a "Fastball Tournament." Starting on the first Saturday of June or right at the end of May, it was time to start off the rotation in Elrose, then White Bear, then Bickleigh and Kyle on the same day and finally the world-recognized Eston Gopher Days on July First. In addition, there would be Sunday tournaments at Eston Riverside Park and Clearwater Lake.

Chuckwagon races *(photo courtesy Roger and Evan Stahl of Bruno, Sask.)*

Folks from all the local communities, including extended family and friends from wherever, would come out to a day of fun, food and festivities. There would be cowboy and cowgirl types who would bring out their well trained and groomed horses for the light horse show, complete with fancy saddles, belt buckles, hats and boots. Then there were the 4-H kids who showed off their cattle, which they had raised and groomed from the year before. One section of the sports ground would welcome this fraternity of cattle and horse-lovers who arrived with their trucks, and horse or cattle trailers, complete with bales of hay and straw for the nutrition and comfort of their beloved animals. That area quickly grew into a mini-western town with house trailers and tents of all ages and descriptions. One of the trailers soon took on the duties as the saloon and meeting place. Participants in both the horse and cattle competitions sported large red and blue ribbons, which they had won. In the evening, the horse and cattle side of the sports ground capped the competitions off with the loud, dusty, and dangerous chuckwagon races. They began at the sound of a roaring horn and were followed by the descriptive commentary of a knowledgeable announcer on the "PA system." No driver I ever saw still had his cowboy hat on by the end of the race.

For me, as a self-proclaimed ball playing fanatic, I loved the ball tournaments most of all. That was when you really felt like a team member, playing above your natural skill level to represent your town and area. We never knew our calibre of play except to be compared by whether or not we could win against our sister towns. As a fastball pitcher of some success, I absolutely loved the swish sound when I would throw a ball with a sweeping curve, and an accomplished batter would swing as hard as he could, but missed it by what looked like a mile. When you are the pitcher on a good day, you can be on top of the world, but on bad days you can be the goat. That was when the Coach needed to be quick with your replacement, even if you thought you could pull your socks up. We were proud of our blue and gold Lacadena Laggers ball sweaters that we wore with our regular blue jeans, and we admired how folks from Lacadena and Tyner would gather around to cheer us on. As trite and egocentric as it sounds, we felt like local heroes carrying our local flag to victory. I think we always won a few more than we lost.

The Eston Sports Day / Gopher Derby was one of the events which happened on July 1—Canada Day. What made it somewhat unique from other "Sports Days" was the full-fledged betting opportunity on races performed by actual live home grown gophers snared and recruited from down by the hills near "Goose River Lodge." The organization committee built a vertical set of tracks one on top of each other like a ten story high rise of long narrow condos with glass fronts. Through each level of the clear glass you could cheer on your betted gopher as he eagerly or sometimes reluctantly scampered from the starting gate on the left side to the finish line into a hole on the right side. This was an obvious centre of attraction for young and old, also catching the attention of the media from hundreds of miles away. After the event, the playful creatures would be safely returned to the prairie wild, back with the cactus blossoms and the sagebrush. The soft sand on the river hills was perfect for digging in their new homes.

Eston concrete gopher and Pioneer elevators
(photo courtesy of Eston Museum)

The Gopher Race Track *(photo courtesy Eston Museum)*

Tyner Memorial Hall—an integrated facility

The Harvest Moon Ball In Tyner Hall

Our local mens' fastball team—the Lacadena Laggers made up of players from Tyner and Lacadena—brought the first metropolitan fashion the "dine and dance" to our community and planted the cultural seeds for its later imposition on many other rural towns for miles around. We were ahead of the pack. Our local community dances in Tyner, or anywhere else in the region before that time, did not serve booze inside the premises. However, the desired liquid courage was always available outside the hall under the blanket of darkness. Sometime in the early 1970's our town hall lost some of its purity and in the view of some local residents it came of age. Fred, a Tyner United Church Elder and dedicated "Hall Caretaker," said our hall lost her virginity to a bunch of local ball players at "The Harvest Moon Ball." Incidentally, Fred, after more than considerable persuasion, perhaps unwittingly had given his permission for the event. His respected blessing was needed to make it happen.

As with most incremental changes in life, this new venture came from one person's practical idea backed up by a few supporters with the courage to make it happen. In this instance it was our "Coach" who championed the idea that was hashed over with a beer at an informal meeting down in the White Bear Pub, where after getting the team go-ahead it was turned over to the best implementation sub-committee possible. The Coach, Joe Pool the Pool elevator man and I were duly elected to form that progressive but boisterous sub-committee. We were told to make it happen. I guess the real question in the minds of our ball team colleagues was "How much trouble can these guys get into between now and the date of the Harvest Moon Ball?" Could this ad-hoc implementation committee even get it done?

There was good reason for concern because in those days having as much fun as possible, by taking as little responsibility as possible, was our creed. Add in our love for fast cars and serious drinking and you could come up with the prediction that a disaster was ready to happen. Nonetheless this whole idea came from our coach, who like Martin Luther King Junior or Walt Disney, had a dream or a vision and seemed to know what he wanted. Joe and I were along for the ride to provide moral support and help implement our new social policy. As our team mission, we set out to prove our Lacadena-Laggers Fastball Team could, as a money-making function, put on the very first dine and dance to be held anywhere in our local territory.

The first task in our work plan would be a fun-filled trip to Eston, about twenty-five miles away, to obtain a liquor permit and the many boxes of assorted booze we would sell at the dance. Our fearless leader, the coach, pumped by our progress so far, decided that we would take his new car but that Joe would drive. Joe was a big guy who liked to have a drink or two, as we all did, but it was very hard to tell when he was drinking or when he had a few more than needed. I guess we felt safe in his work worn hands. Much of our day in Eston was spent personally advertising the upcoming Harvest Moon Ball to anyone we met or tracked down.

On our trip home from this very successful liquor run we decided to reward ourselves with a few drinks on the road. Strangely enough in those days, the law didn't worry about how inebriated the driver was so long as there was no open liquor in the vehicle. Fortunately for us, Joe Pool could drink an awful lot before it seemed to affect his driving ability, at least so we thought. The police officer who stopped us out on the grid road, south of Eston, seemed to agree with our confidence in Joe, because he just said

"Have a good evening" and let us go. It was time to head home, Joe and I to Tyner where we lived, and the coach would take over to drive his own car the rest of the way home to Lacadena. First, we needed to unload the dozen or so boxes of alcohol in Tyner, ready for the "Harvest Moon Ball" on that Saturday. Everything had gone as smooth as silk—so far.

The next day began with a call from the coach to both Joe and I. Apparently the night before, after he let us off in Tyner our fearless leader may have been a little bit more affected by our day of fun than we thought, and his ability to drive might have been more impaired than was safe. On Main Street in Lacadena, right next to the landmark Chinese Café, there is a great big power pole right by the sidewalk on the corner, where the adjacent cement sidewalks meet. On the way around that corner, our coach's brand new, long wheelbase, grey-blue Buick car managed to end up unceremoniously wrapped around that big power pole. The car was totalled off, but nobody was hurt, except for our coach's pride of ownership. However, the large extended Chinese family living in the cafe caused a terrible uproar as they ran for their lives, as any of us would do. Sparks were flying off the black power transformer near the top of that pole, which shook from the blow and appeared ready to fall. The coach's wife, who was nestled in bed at their home just down the street, heard the commotion and soon arrived on the scene to see what had happened. She was obviously not impressed to find her usually dear husband involved with this looming disaster. She is a very nice, generally calm and cool lady of much beauty, who could be heard for miles around as she evaluated our coach's performance. Some say it was the loudest performance review that anybody from near and far had ever heard. Our coach had some explaining to do.

Not deterred, the next day it was business as usual for our implementation committee. However, the first thing we needed to do was pick up the coach and take him to Eston to get his old car, which had been sitting on a consignment sales lot since he got his new but now twisted Buick. It would need to be licensed and registered for its unexpected return to service. We needed to get back to the work of our implementation committee, because Saturday night would be the night of the first ever "Harvest Moon Ball" at the Tyner Memorial Hall.

This was not going to be a traditional "dance," this was going to be a "ball." People would be sitting around decorated tables with bright baseball designed tablecloths, and drinks would be served within the building, not out in the parking lot. The guys, in unusual fashions for young ballplayers would be semi-formal in attire with dress pants and finely knit pullover sweaters. The ladies, however, whether wives or girlfriends or just friends, would go all out to dress in the most revealing and contemporary gowns that most God-fearing farmers had ever seen. Earthy autumn colours were the rage but solid-black to too-hot red mini-dresses also joined the fashion party. Some of the dresses, with the generous exposure of skin from cleavage, legs, and arms, would be famous for years to come. The ladies' shoes had heels like nobody had seen or endured before. Hairdos would be planned and executed in the most delightful styles. Then there were the glorious perfumes in the air, which seemed to stimulate thoughts of sexual games unlike anything this innocent rural community hall had ever seen before. That romantic and sexy atmosphere complemented with smooth drink was about to inspire attractive people to take some risky chances, which could put marriages in danger and single folks in trouble with married folks and vice-versa. There was something about that dance, with the music, the movement, the whispering, the touching, the variety of drink, and the relaxation, which in some instances temporarily removed the limitations of a marriage certificate. Alcohol and hormones would soon take control of the event.

When we said the Harvest Moon Ball would be dine and dance, we really meant it. Dine we did. The wives and girlfriends took very seriously the task of providing the delicious midnight meal. It would be a full course turkey dinner, much like you would tackle on Christmas Day. Remember, this event was after harvest in October when the fruits of our farm and garden work were abundant. Several large turkeys were roasted, dressing was inserted, a bushel or more of new potatoes were cooked and mashed, vegetables of all sorts were prepared along with salads of all kinds. Then there were cakes, pies, fruit dishes, cookies and ice cream, of almost every flavour.

People came from all of the competing teams in the ball league and from other places from all around. The hall was filled fuller than anyone could imagine. The dress code had been well publicized and the glamour of metropolitan dress and demeanour was evident throughout the night. The motto for the occasion seemed to be "If you've got it, then flaunt it." The live band played all the Oldie Goldies and Big Band dance renditions. It was up to our Lacadena Lagger fastball team to pitch in (pardon the pun) to keep the peace and do the work necessary to pull this event off. I must say, we did a great job, as would be expected from a well-oiled team. That is not to suggest that we didn't enjoy ourselves as much or more than anyone else. Despite his recent traffic misfortune, the coach was in a joyous mood and was extremely proud of his team. His beautiful and kind wife had also recovered from the coach's misfortune. She also had great fun and seemed very proud of our coach and what he had done for the team.

After the music stopped and the dance was over at about two in the morning, it was time for the team and close friends to clean up the hall, do the dishes, pack the leftovers, close the bar and count the cash. That process would mean more relaxing drinks, more food and great conversation for the insiders closest to team members. There is something about team participation that generates meaningful friendships.

The last administrative duty for our implementation committee after the "ball" was to count the cash. After all costs were deducted, we had made a profit of six hundred and fifty-nine dollars and seventy-five cents. As a committee, we recorded and immediately reported that number to the other team members present. We also wrote it down in a do-it-yourself financial statement, because records are important. The total financial balance was recorded with a marker pen on a paper towel and stuck into a plastic bread bag with the money. A large cheer went up in the now clean and restored Tyner Hall as our coach took charge of that sack of hard earned money.

After most people, including the team members as well as Joe Pool and myself thought this historic occasion was finished, our flamboyant coach disappeared with our profits in hand. Where could he be? Much to our dismay, we later discovered that he was thought to be on an impromptu "toot" with some of the Elrose fastball team members. Adding even more to our growing concern was the fact that as we had suspected, the coach had our cash profit from the Harvest Moon Ball in his possession. We felt partly responsible for what appeared to be some unethical and costly actions by our dedicated coach. To be very honest, I did not know what to do, but I didn't want to believe that our leader-mentor was up to no good, and that he had double-crossed us and everyone else on our team. The brutal facts had me very worried. What made a bad story even worse was that the coach's wife didn't know where he was either.

Then on Wednesday afternoon, the replacement car the coach had been driving was seen parked back at his home in Lacadena. Someone called Joe right away with the news and he picked me up, as quickly as he could. We needed to confront our coach to hear his explanation. Therefore, we set off to Lacadena with a high level of scorn and sadness in the back of our critical minds.

On our arrival in Lacadena, I guess I was the one who asked the first question when we finally encountered the now sheepish looking, very tired and obviously hung-over coach. My probing question was naturally very short and to the point: "Coach—where the hell have you been?" Without a word, but with a characteristic sly smile, his first response to me was to reach into his pocket without a word and pull out a ragged-looking plastic bag and hand it to me. Without hesitation, in front of Joe and the coach's flustered and also sceptical wife, I counted out all of the profits from the Harvest Moon Ball, right down to the last cent in what looked to me to be the exact same bills we previously counted that last Saturday night. Smiles began to flow all around as we knew nothing had gone wrong with our innovative venture. Our usually trusted mentor had not let us down financially or morally as some—myself included—may have suspected.

Our fearless coach then related how without much thought about fiduciary appearances, he had jumped in with some of the Elrose guys who needed to deliver a truck one of them sold to someone up in Prince Albert. He said; "the Elrose truck dealer offered our Lacadena team an extra hundred dollars if I drove another of his trucks up to bring him home." Our visionary leader then before our curious eyes transferred another one hundred dollars into the plastic bag that held our team coffers. Very quickly Joe and I made sure the team and the community supporters knew that our coach and our implementation team had done a good job, right down to the last scary detail. To this very day, many of our fastball team members, including myself, still have our royal blue team jackets we all received with some of the money raised, and we also have the fond memories of a fun time at the Harvest Moon Ball. We had prepared the way for many joyful "dine and dances" in the future.

The White Bear Pub

The White Bear Welcoming Polar Bear

Nobody I know has ever seen a bear, never mind a polar bear, in flat-prairie west central Saskatchewan. But then again nobody I knew ever saw a burning bush or the parting of the Red Sea. Never the less the prosperous little community of White Bear is named after a polar bear. I like to think that an inquisitive old bear, once strayed down from the grain terminal at Churchill Manitoba, to see where all that high quality durum wheat came from.

Jokingly, I once said that the closest thing we have to a university within twenty-five miles of my hometown is the White Bear Hotel. I went on to suggest that by visiting our world famous pub, you could expect to gain an education beyond your schooling. I refer to it as getting clever. You might also affectionately describe our pub as the local board of education. This rather unique yet simplistic establishment has, for many years now, attracted patrons from Kyle, White Bear, Elrose, Eston, Plato, Isham, Tyner, Lacadena, Clearwater Lake, and Saskatchewan Landing, as well as many other more distant locations from time to time. It has a provincial reputation, which is continually expanding by word of mouth.

For many folks in our home area, a trip to the White Bear Pub is a routine but exciting occasion often viewed as a just reward for a few days of hard farm work. For others, it is a quick stop-off place to pick up some beer for hot days at home on the deck. However, some folks, free from the daily chores of employment use it as a second home or a hang out. Nonetheless, the White Bear Pub has evolved into an iconic destination spot that regularly brings together a number of locals from our surrounding farms and towns for relaxation and fun. Not to mention the untarnished wisdom that spills over from our version of "rural academia."

Leaving my hometown of Tyner, you would follow south and east along the thin worn-out oil road, across the remainder of the Tyner Flat through Lacadena, and up the steep hill into the sleepy little town of White Bear. It sits right on top of a long and narrow coulee which has been dammed up to retain a reasonable amount of water into a small lake—the White Bear Dam—where with patience and the right hook you can catch a pan sized rainbow trout. It is about a sixteen-mile trip from Tyner to White Bear, taking about twenty minutes—depending on your thirst for cool drink and deep conversation.

My first impressions of this rustic old pub, as a young boy, were not terribly pleasant. However, when I think about it, perhaps my evaluation may have been far too harsh, jaded by my unfortunate circumstance. As it happened, our farm neighbours near our town were frequent visitors to all the bars in the area, often for greatly extended hours during the day and night. As a result, after school my parents often looked after their son, who was about my age and went to my school in Tyner. Often that meant hastily organized overnight stays for him at our house, because his folks would not get home at a reasonable hour. In return for that regular support provided by my parents, this wayward couple with good intentions but bad habits would occasionally invite me to travel with them and their son to places like Clearwater Lake for a swim. I think their intentions were honourable, but the problem with their plan was that White Bear Pub was right on the usual road to Clearwater Lake. So, that promised trip to the lake for a swim often amounted to us boys, at about twelve or so years of age, finding ourselves confined to a vehicle sitting in front of a pub for hours at a time. I must confess that I began to develop a growing hatred for the mundane outward appearance of the White Bear Hotel for that reason.

White Bear Hotel

More than a few times, we spent three or four hours outside the pub with pop and snacks delivered part way through our experience, in a futile attempt to eliminate our boredom. Unfortunately, I got very familiar with the old, dull white, nearly grey, weather-worn, cedar-siding barely hanging onto the gloomy exterior of that popular establishment, long before I could go inside. Please don't ask me to describe the complications of later going to the lake for a short swim well after dark with a few very heavy-set, uncoordinated adults who had been in the bar for most of the day. That would be my day of pay-back entertainment.

Even on the best day, it is not a very classy welcome when you enter through the shabby, old, unpainted and uninviting hotel door. You can go left up the stairs to a few usually vacant and small hotel rooms, or walk straight into the hotel lobby and kitchen area, or turn right going through an interior door heading into the pub. Right where you make that choice, there is a bulletin board letting you know who has what for sale, and what community events such as the Kyle rodeo are coming up. Once inside the bar there is a wide-screen TV close to the entry, which you can watch from the long narrow arrangement of bar tables, chairs, and benches along the wall, as well as from a few cafe style booths that have been added more recently. As you move further toward the front of the bar room, on your left you go past two ancient but clean bathrooms and a do-it-yourself steak pit. Right then you finally arrive at the counter and cash register, by the huge walk-in cooler. The walls all around are covered with an abundance of pictures, usually popular artist prints of majestic animals, hanging below a long upper shelf with a number of antique radios and such sitting on it. At any season, Christmas-style lights hang almost everywhere to give a coloured starlight effect. You really need to see it to believe it and to feel it. The air conditioner keeps the place quite cool as an attractive oasis on hot summer days, while hard liquor does the same by warming your heart on cold winter nights.

The unofficial uniform the male bartenders have worn over the years consists of black dress pants, a white dress shirt with long sleeves rolled up to the elbows, and the top two or three shirt buttons undone to expose some manly chest hair. The bartender needs to be part preacher, teacher, social worker and entrepreneur—not necessarily in that order. Unfortunately, for business reasons the scope and depth of wisdom dispensed by these gentlemen, by necessity, needs to be a dull encapsulation of the "clientele group-think" from that day or week. A bartender generating conflict was never good for that type of business. The wise-guy, know-it-all, conflict-inspiring role would be better played by one or more of the guests. Nonetheless, a tactical bartender, I guess to escape total boredom, would sometimes bring up controversial issues from the news or elsewhere to generate some heated discussion. In total, the bartender considered his role as one of boss and manager, with the freedom to get away on business in order to pick up supplies, or visit the credit union or bank.

Much of the punch-the-clock, routine, day-in and day-out work fell on the shoulders of the bartender's wife and business partner along with some part-time female help for busier times. Working in the often-hectic kitchen was almost always the job of the wife and her helpers. Somebody needed to be there from opening until closing time, which went late into the night before the activities of the day were complete. The next day would be much like the day before and the day after that again. They would not let on, but the same faces of the same people, in the same chairs with the same problems, hearing the same ineffective solutions, from people who often let liquor do their talking, must have been less than enjoyable for the bartending staff. I guess it was all about the money.

During goose and duck hunting season, and to a lesser extent in deer season, hunters from places like Texas, Minnesota, Indiana or almost anywhere else would come to stay at the hotel to experience the quaint environment of rural Saskatchewan and hunt for sport, not for food. The eight or so high and narrow rooms upstairs that shared a bathroom including a shower, would come to life for a couple months each year from the middle of September on into November. Friendships were made throughout the district as hunters staying at the hotel sought out the farmers' or ranchers' permission to hunt on their land. Since the harvest was over and the crop was finally safely in the bin, the farmers and ranchers had more time for relaxation and socializing in hunting season before curling and hockey began. The White Bear Pub would adjust for all of those routine changes of the seasons.

What I do know is that on special occasions, with relatively cheerful smiles, almost like in a dance mode, the bartender and staff would swirl around from table to table taking complicated and confusing orders for food and drink with an almost mistake-free record of performance. For certain events like green beer—Saint Patrick's Day—the bartender's wife and her perky young female help, would delight the patrons by shedding what usually was quite ordinary but appropriate garb, in exchange for scanty party attire. Although I was never privy to the exact numbers, I presumed from my own animal instincts and my own careless actions, that there was a direct relationship between the tips they collected and the abundance of bosom cleavage and bare leg exposure presented by those friendly ladies. An adventurous gal with a generous gift of personality and endowment, and the ability to bend over or rub the customer at the right time, could easily boost her compensation to several times normal. Some girls will do almost anything to make a living, and sometimes more. As one older fellow explained as he dropped a hefty tip into the waitress' skimpy top explained, "I might be old but I'm not dead yet."

I guess when I was about ten years of age, women gained the right to attend the pubs in Saskatchewan, much to the displeasure of many male traditionalists. Some of the stories going back to that era before women were included seemed to me to be more violent and crude, but I obviously didn't see for myself. My observations about White Bear Pub have always been from after the inclusion of women, and for the most part I've noticed that by their own preference, ladies generally attend mainly in the evenings and on weekends. Men usually make up the regular loyal contingent during weekday activity, and you could be challenged to imagine when some of those guys worked, if they did.

My favourite story involving a lady in the bar was when Peter and Alice joined our table. For some reason we were drilling down into the issue of people getting remarried after a partner has passed away. Peter said to Alice, "If I die before you, I want you to sell all of my stuff, everything." "Why?" Alice asks. "Because you are quite an attractive young-looking women, and you will likely find a new man. I don't want some asshole driving my truck, golfing with my golf clubs, hunting with my guns, or walking Sport, my loyal hunting dog." Alice responds, "What makes you think I would marry another asshole?" We all had a good chuckle about that one but I knew I had seen Peter laugh harder before at less of a joke. I think that message cut deeper than he expected. It became clear to me that women can participate in sarcastic debate, along with anybody else at the White Bear Pub.

Sometimes I prognosticate that one good reason to go to the pub is because the only people in the community who are not being talked about are the people who are there talking about everybody else. The White Bear Pub is not a bakery, so nothing is sugar-coated around there. We tell it like it is and our language is not purified, so the F-Bomb is sometimes a part of the lexicon. The stories are somewhat random, usually embellished and often repeated. It is wise when visiting our pub to accept the old adage that "You can't believe half the lies you hear."

With any luck at all you will arrive at the Pub when the front table is hosting four of the notorious regulars, who have obviously been sent by God himself/herself to solve all the world's problems. Your best bet to encounter these celebrities would be on a rainy day when the work stops and their creative communications skills come alive. The life-skills seminar begins. This gaggle of high priests, each ready to preach their own version of the holy gospel of farm business and rural life includes Windy Bill, Gene the Genius, Len the Lefty, and Sean the Shyster.

For the most part, this gang of "Prairie Wise Guys" is not necessarily guided by valid research, formal education or solid facts. Their wisdom is fuelled by cold beer or hard liquor and self-serving opinions, but these fellows have the capacity to laugh, yell, debate, insult and argue for several hours at a sitting. For those of us who often had the privilege to observe these sessions, we could not understand how people of such diverse philosophies and opinions could spend so much time together. Perhaps it was their unwavering commitment to their own self-interest or political cause, along with a burning desire to wave their flag as high as they can, as often as they are able to. Maybe everyone wanted the last word. I was quite sure they didn't much like each other, but like relatives rather than friends, they were stuck with each other in this rural meeting place, and so they made the best of it.

The only interruptions that are imposed on these verbal Olympians are times for frequent bathroom relief and constant drink refills. Sometimes the action is broken by a telephone call from a concerned wife or family member. One time before cell phones were available, I remember a call came that the bartender answered and said to Bill, "It's your wife Lois on the phone." Bill slyly but cleverly told the bartender to tell her he wasn't there. The loyal but astute bartender returned to the phone and said; "He says he's not here." I guess she was not impressed and Bill was soon at the phone with more than a few "Yes, Dears" trying to dig himself out of a dangerous marital quagmire.

Bill is a stocky fellow who looks like he definitely was designed more for comfort than for speed or work. He has a friendly round face, with dark but silverish hair and tanned complexion. Perhaps he is his own worst enemy, as he talks way more than he listens. Windy Bill is not out to hurt anyone and can be quite generous to the community. Some say he once did his fair share of work in the distant past, but others say he always saw life as more for fun and relaxation. His wife jokes that Bill has never been afraid of work because he can lay down right beside it for hours.

Windy Bill is the cowboy at the table "with the hat and with the cattle," often wearing his big brown ten gallon hat and fine designed, soft leather, brown cowboy boots. Somebody once said that Bill had a two-gallon head in a ten-gallon hat. Anyway, his slick slim-cut designer western shirts about two sizes too small, with the white enamel buttons and a pull-cord cowboy tie, along with tighter than comfortable blue jeans make Bill look something like the cowboy he had once been. However, his duds clearly have shrunk over time in the damp country air. Story has it that Bill was a pretty good "bronc-rider" and "bull-rider" in his day, even with an entry at the Calgary Stampede where he had a good performance. He had also served as the "pick-up" man at the local Clearwater Lake Rodeo held each year for many years over near the lake. That was a unique experience where the rodeo corrals were set up in the valley so the audience could park their vehicles on the adjacent side hills to view the events from a blanket beside their cars and trucks. Bill—in true cowboy tradition—loves to square dance, and he evolved into a very colourful and qualified "caller" for the local dances. "Now greet your partner and swing her to the right-doesie-doe and around you go, make a star and don't be slow" he would loudly sing to the dancers as the western-swing music played. "Grab those heifers and spin them around"—"Bow to your partner and head for town"—he would continue.

Something else Bill was famous for was when his family would host one of the local branding parties, where neighbours and friends from miles around in mid-summer would help with the branding chores for the new spring calves that had arrived in March. It was a day or two of hard work, ending with a feast and a cowboy party complete with some amateur cowboy poetry. The branding also provided the groceries for the annual "Prairie Oyster Feed" over at the Clearwater Lake Hotel.

When everyone else gets nine tenths of an inch of rain, or none at all, Bill gets two inches, probably with fifty percent hail damage that nobody else got. When the durum crop on average in the area goes thirty-five bushels to the acre, then Bill gets forty-three bushels to the acre. In the busy seeding and harvest seasons, if the best of the best workers log in eighteen-hour work days, then Windy Bill claims he just put in nine twenty-one hour days. Should anybody make thirty-five miles to the gallon with their pickup truck, Bill personally calculates that his new Ford got forty-one miles to the gallon on his trip to Swift Current yesterday. Even his own family members say they find his embellishment of the truth irritating. Bill embarrassed his wife when he told the group "The town of White Bear has never grown over the years because every time a baby was born, a man left town."

No farm or ranch could exist without a trusty rain gauge located strategically on an unobstructed post within close proximity to the house. Its recorded testimony would be the source of talking points at the outset of any conversation soon after a rain. Everybody knew of Windy Bills temptation to embellish his fortunes when it came to rain in his gauge. Therefore from time to time, he was absolutely correct when he saw more water in his gauge then was in all his neighbours' gauges put together. That was because his friends made sure his gauge contained some extra bladder-driven yellow rain. Bill never caught on, or else played along.

One time I was told that after much drink, a real family feud broke out when one of Windy Bill's usually loyal cousins, at a special birthday party after one of Bill's tall tales, tried to accuse Bill of being full of BS. It made Bill so angry that he took his 12-gauge shotgun under his arm and headed out the door, telling his family that he was going to shoot himself because "I get no respect around here." Nobody followed him out or tried to stop him as he surely had hoped, but there was some worry in the farmhouse air. Pretty soon a loud muffled gunshot rang out. Still, nobody moved to check it out. What had Bill gone and done to himself? The tension level in the house quickly rose and some serious faces considered the brutal possibilities, as folks seriously wondered what Bill had finally done. Then a second loud unmistakeable shotgun blast exploded into the sky, followed by yet a third shot. Then every family member, in the grand old farm home, relaxed and repeated in unison "What a bullshitter," as Windy Bill came in with some Hungarian Partridges he had just bagged.

Anyway, Bill is not much interested in the political talk of the day because he comes from the lighter side of life. His excuse for not learning about or participating in the political process is because as he says, "Politicians are all the same and they are all a bunch of crooks." He often says there are an awful lot of improvements that need to be made in the democratic system before he would lower himself to get involved. The last thing he wants to do is waste his valuable time attending meetings. What he seems to be saying is that if everyone else who is involved fixes everything up to where it suits his fancy, he may be willing to participate and do his part. Until then, for Bill, democracy is someone else's responsibility.

Then there is Gene the Genius, who always occupies a seat at the front table. Gene is an ordinary-looking fellow of medium to small stature with a fair complexion and a receding brownish hairline. His face has a natural grin, which to the observer appears to represent someone who feels a bit superior to others. I think you would say he presents a rather cocky and confident façade. Gene is the poster boy for free enterprise gone wild. He has very little concern for anyone who is down on their luck and doesn't have the wealth to participate in the sacred marketplace Gene worships. His economic religion is grounded in the user-pay theory, which says if you want something you need to pay for it. If you don't have wealth, then you don't deserve the services. He hates paying taxes and he condemns governments that collect them. He says governments are inefficient and corrupt, so they should not regulate or stimulate the economy or provide support payments to anyone, even the poor. Gene is in the choir that sings all the gospel hymns of deregulation, privatization, self-determination and personal greed. He knows the poor will always be with us because they are lazy. For Gene, helping kids from poor families is a travesty of justice, because he knows of specific examples where some parents have smoked, drank and gambled all their money away. As if it was the kid's fault.

Gene tells the crowd, "Who said life should be fair? We get what we work for," even though that motto has not been the reality for Gene himself. He is well-placed to succeed in the market he loves and worships because of how much better he was set up compared to other genuine self-made farmers. He doesn't even need to take out crop production insurance or put on hail insurance. As one neighbour has stated, "He was so well set up on that farm that it will take three generations for his clan to screw it up." It is easy to become envious of Gene and how he was given the best of everything. However, as he gloats about his status to everyone inside the bar or elsewhere, he soon grates on your nerves and your better judgement.

One example of his arrogance is when he tries to impress people by pulling out of his pocket a mitt-full of one-hundred-dollar bills just to pay for one or two drinks.

There is no doubt that Gene is a farmer who sees himself as more in line with giant multi-national business owners and management, than part of the small business working class. He dislikes unions with a passion, especially the leaders or "bosses" that he thinks run them. He won't acknowledge that the union leaders are democratically elected by the membership. Workers are too greedy, he purports, and strikes or other labour movement strategies to improve the lives of their members should be outlawed. Minimum wages, in his opinion, are breaking the country and unemployment is a disgrace because there are lots of jobs out there if people were not just too damn lazy to work. He says it is time for the unemployed to get off their asses and go to work.

Gene is totally comfortable and financially secure to the point of aggressive smugness, given the farm operation he was set up with by his parents and a frugal bachelor uncle. Armed with his fortunate unearned economic power, which provides his ability to participate fully in the market economy, Gene has grown to admire what he calls the "free market." He will not admit that it is not free to everyone because it is only available to those with wealth. He has no concern for anyone with less wealth than he has, especially anyone coming from aboriginal or Asian heritage. More recently, Gene has become famous as spokesman for a rump group that campaigned for the elimination of the Canadian Wheat Board's single-desk selling monopoly. In that instance, Gene foolishly fought in the name of free-enterprise for the freedom to earn less as a farmer.

Gene's family was very active in the "KOD" keep our doctors movement helping Saskatchewan doctors resist the implementation of government-paid medical care insurance in the early 1960's. He hates the CBC because he sees it as a mouthpiece for socialist propaganda. Gene condemns any form of liberal arts education, viewing it as scandalous and wasteful. All education he feels should be directed specifically at the technical trades. Moving jobs overseas, to use cheap labour in return for lower purchase prices is fine with him. He takes the position that corporations should be trusted to police or regulate themselves, without government interference. Gene strongly believes that what is good for business is always better for the country. According to Gene the privileges and the pomp and ceremony, associated with the monarchy and the Canadian Senate, are worthwhile costs to provide a just reward to the elite in our society. An elite he sees himself aligned with.

If you want to see a grown man flip out into a state of complete irrationality and dangerous convulsions, just mention that some form of gun control is needed. Tell Gene that it is needed to protect our society from military style automatic weapons beyond those needed for sport hunting. He loves his guns more than almost anything else. He has been lured into that "War is a harmless game" mentality, where he believes you can bomb your way to peace, imposing your values onto the rest of the world and then come home unharmed and put your feet up. Then, he brushes off with disgust any thought that wars may be part of a hidden agenda to feed the profit motive of the military supply establishment. If war is to take over foreign oil production, that is fine with him. The destruction of life, limb and mind in foreign wars is not a concern he cares much about. Gene knows that poor families who need the money will always be their to answer the call to war whether our governments take care of our veterans or not. The wealthy can depend on them.

Gene can't understand why anyone would lament the loss of our romantic community spirit from the past. He would say "Don't talk to me about the past because I don't live there anymore." In his view, all the land will get farmed and all the supplies and services will be purchased, even if the corporate farms make the decisions and business is done by volume purchases in the city. Gene says, "I will go as far as I have to go to get the best deal." When you farm as big as I do, you can demand special treatment. The last thing we need around here is any government assistance for smaller family farms because have you ever seen anything the government could run efficiently? Co-ops or any other form of working together is not necessary as far as I am concerned. Those old stories about farmers and communities coming together to raise a barn, build a rink, or help a neighbour suffering from an illness in my opinion are just fairy tales from the past." His favourite saying about those who wish to support the struggling communities is "When the horse is dead dismount." Gene says we must move on and not obstruct progress. Progress as he defines it, in his own image.

Taking the seat across the table from Gene is Len the Lefty, who is now a well-established mid-sized farmer, and who unlike Gene, started out with very little family support. He combined hard work with a broad education and useful government support programs to help his family farm succeed. Much of the credit goes to his hard-working wife, who along with him, held down several off-farm jobs necessary to subsidize their farm. Len is a large friendly man in great physical shape but he is starting to show his physical age, even though he certainly has kept his sharp mind with his constant reading and research. With a well-groomed reddish-grey beard and longer than normal wavy brown and silver hair, he has a professorial appearance.

I think it would be safe to say that Len has very little patience with folks like Gene who claim to know everything. Gene once bragged that all he started farming with was a sack of wrenches, but he knew just when to enter the industry and buy the land and machinery he required. Len says that Gene's self-promoting fairy tales can be ranked with the guy who says he was born in the outback in a cabin he built himself. Or maybe with the guy who claimed he was born when he was twelve years old and had to go out and work to afford a mother. Len submits that, generally speaking, all farmers got a start with various amounts of help from family, neighbours, banks, co-operatives or government programs. He refutes Gene's rationale when Gene glorifies the trickle-down theory of economics, believing that making the rich richer will mean more wealth will trickle down to the poor. In reply, Len often mentions what he calls "the ovarian lottery," which he says is just being lucky enough to be born into a family of wealth. Len maintains that for that reason alone, a person can get a larger share of resources and services than others not born so lucky. Len boldly tells Gene that he was born into the privileged elite and his claim to be a self-made man is craziness.

Len has immense respect for the lessons of history. His dad, a decorated WWII veteran, had always told Len how real and brutal all wars are, no matter how justified the reasons may seem to be for the war. Len says we should work to avoid war and we certainly should not glorify it. He says that if in the slim chance that war is necessary, we have a sacred obligation to care for our veterans.

Based on his own study of history Len tells the guys: "Just because my grandfather didn't have a computer doesn't mean he was stupid." Len has read much of the literature about the farmer driven "Progressive Movement" responding to concerns about the unfair treatment farmers historically faced from the corporate monopolies organized by the railroads, the banks, the grain companies, machine companies, and chemical or seed companies. He knows how farmers were cheated on grades, dockage, weights and prices by the grain monopolies. Unlike Gene, Len thinks it is unwise to put the fox in charge of the hen house by allowing corporate monopolies to exercise self-discipline in the deregulated market they request. Len supports the market economy, but maintains that it needs to be regulated in order to provide fairness for players who hold less market power because of less wealth. On the farm input side, he says the corporate monopolies are able to set the level of input costs the farmer must pay. However, that cost level is not a direct function of farm product prices that determine farm revenues. Hence, there can be a damaging "cost-price squeeze" when the prices farmers receive are down when farm input costs are pushed up by monopoly pressures. That market dysfunction is squeezing more and more family farms out of business and depriving our towns of the population they need to survive.

He preaches that farmers need their own monopolies in order to compete. Len says farmer controlled monopolies such as the Canadian Wheat Board should be used to level the playing field for farmers in the global market. He is also a vigorous supporter of crown corporations, which he says are needed to keep local control of our resources and utility networks. Similar regulations, in his view, are needed to protect the integrity of our communications and news networks. Farmers, he says, will learn the hard way that the true fight for farm survival is not with each other, and it never has been. Len thinks they will come to understand that as individual farms, no matter their size, they will not be able to buy their way to the head of the line to receive preference and favouritism that some companies falsely promise them. That false sense of individual security that some better-off farmers cherish will soon vanish as agribusiness strips farmers collectively of more of their share of the growing food chain wealth. He hopes farmers will learn that what is good for the general food economy and agribusiness in particular is not always good for farmers. Len says the farmers' share of the food business wealth has been going down, especially since the elimination of the CWB single desk marketing.

Len wants all Saskatchewan people, including rural folks, to benefit fully from the rational exploitation and development of our resources such as potash, oil, gas, uranium and agricultural products. Royalties, rents and taxes need to be fair but at sufficient levels to allow municipal, provincial and federal governments the financial capacity to provide the public goods and services needed such as health care, education, and environmental protection. In his view, selling off our raw resources to be value-added at some other location with lost manufacturing and employment opportunities is not fair for our people. He says we owe it to ourselves to demand more. To him, selling off our resources at low royalties, without amortizing the profits for future generations, is absurd, especially if it destroys our environment. Len likes to cite Norway as a country that has been saving the profits from depleting natural resources for the future. They have a heritage fund worth a million dollars per citizen from oil revenues and we are in debt. "Maybe we should eat more ludefiske and lefsa like the Thompsons and Heslas over at Kyle," he declares.

Len credits unions with obtaining better wages and benefits for working people. He doesn't understand why non-union folks condemn unions for gaining higher compensation for it's workers, instead of forming or joining a union to get the same wages and benefits for themselves.

Len the Lefty asks his friends at the front table why governments, oil and gas companies, and pipeline companies want to send unrefined heavy oil through a new pipeline network to provide manufacturing and refining jobs in other countries such as the USA, when it could be done here. He informs them that the pressure needed in the line to send refined oil would be much less than shipping the heavy crude. That fact alone would significantly reduce the danger of pipeline ruptures and oil spills that destroy our environment. As a strong proponent of co-operatives, Len points to the Co-operative Refinery in Regina as an example of how capital and management can be put together to do what is needed right here at home. Although he cautions that Federated Co-op has drifted away from the democratic co-operative principles and has marginalized local co-op boards and the members. He is concerned and says "if FCL is not taken back by its members it will lose its value as a co-op by merely emulating other profit seeking businesses."

Len pontificates that because they are starved for capital investment and don't want to collect reasonable levels of taxation, even from our natural resources, governments are now putting their faith in the same monopolies which previous governments and co-operatives rose up to protect farmers from. He says governments in a reckless attempt to attract capital to our domestic market are now prepared to even give up national sovereignty and environmental protection in the fine print of devastating global trade agreements. Agreements which allow corporations to litigate for damages against a sovereign nation, should that nation's government decide to regulate an industry in its citizens best interests, if that law reduces profits for the foreign corporation. This craving for capital from both domestic and international sources is also pitting the economy against the environment. There are vast amounts of wealth generated from the market that is used to refuel the vehicles on this destructive path. Len says that wealth is used to lobby governments to deregulate and relax environmental controls while also producing a multi-million dollar corporate spin, falsely convincing people that there is nothing to worry about, and nothing needs to be done about it. Then, some of the wealth is used to lobby and buy the politicians so they won't obstruct the corporate road to environmental destruction.

Of all the four regulars at the front table, without a doubt the most devious and vindictive person is Sean the Shyster, who, by his actions in life has put hypocrisy into motion. He quite eloquently says one thing, but does something else. Over at his seed cleaning plant on his farm, he holds court in his self-defined business coffee shop, where he preaches from his dusty pulpit to his local parish of unsuspecting and somewhat naive farmer friends. He preaches a form of self-serving socialism. He offers free public advice for all on various issues, but always with a thinly disguised hidden agenda expected to result in some form of personal reward at the expense of others. Sean is a rather handsome guy about six feet tall, in healthy physical shape, with a mat of greyish-black curly hair. He is the youngest one at the table, and sees himself as still quite suave and young at heart. He would be ready to cheat his own mother.

The Shyster is known to have gained much of his land base by some form of manipulation or deception, although, he, too, got well set up by his parents. His dad acquired a fair bit of land when many other young men his age in the district, were off to war to fight for our country. His dad had it rather easy to find land because there was a shortage of men left in our country at that time. Sean has added to his free land base from his parents by suddenly pouring on the kindness by wining and dining vulnerable farmers seen as ready to quit farming due to age, illness or family issues. This guy has a knack for turning up at those vulnerable times, without any former interest in the welfare of his targeted farm owner. Then, after he is able to manipulate them into selling him their land at a desperation price, that sudden burst of personal interest in their welfare quickly disappears.

My friend relates the story of how Sean would preach to his captive audience at his seed cleaning plant, about how impossible it would be to pay for land at the current market price. If a particular parcel of land for sale was under review by other members of the group, especially someone who needed more land, the Shyster would explain in precise detail how stupid one would be to pay that expected price for the land. Then as Sean had planned, that interested farmer, based on a sanctimonious lecture from Sean would not muster the courage to make an offer to purchase the land. He would not risk the scorn from Sean a self-professed expert and leader of the pack, if he bid on it. Others in the group would not bid either because they felt the other farmer should get the land. As my friend said, "before the seed plant coffee crew would be home that same day, the Shyster contrary to his own advice, would have bought the land in question." His unfounded manipulative advice had reduced his competition for his own reward. Sean may have forfeited a friend but he got the land. He always had to come first.

Sean doesn't like to talk about it, but there is a story behind why he lost some land he was renting on a crop share agreement from an elderly widow now living in town. In his crop share lease agreement, Sean was on his honour to make sure the widow got paid for her full one-third share of the crop when it was delivered and sold. Story has it that the widow smelled a rat when checking her cash receipts for her grain. They didn't seem to add up to what she and her son from Calgary thought matched the amount of crop grown. One of Sean's neighbours also got suspicious when he noticed a couple of times when he was driving by Sean's place, Sean brought his semi-truck home from harvesting at the rented farm with grain on it that he emptied into his own bin. Why was he bringing grain home to his own farm from the rented land? The neighbour's conscience bothered him to the point where he called the suspicious widow and her son. That information and her audit of the books resulted in her terminating the rental deal with Sean.

Sean had a bachelor uncle with a reasonable amount of wealth, including some land. As a long time dedicated member of the Masons, Eastern Star and the Shriners, that uncle decided to will his estate to the Shriners' Childrens' Hospital. His will had long been in place to do just that. Sean had not been much of a friend or supporter of his uncle up until the uncle suffered a heart attack and became quite ill. That is when Sean quietly went with an unethical lawyer to his uncle, who was in a fragile state of mind, in order to convince him to sign his will over to Sean. Within a short while, Sean had his uncles land and the Shriners' Hospital was left out.

Sean is one of those guys who always has a great idea for a "work bee" over at his place, where he would like to have nine or ten guys help him out with a major project. He will supply a couple of boxes of beer. Sean is not ashamed to ask the guys to bring a tractor and front-end loader, a couple of trucks and numerous tools he doesn't have. If anybody has any leftover building materials, they are welcome to bring them over because he can put them to good use. Not bad all for the price of a few cases of beer. Unfortunately for Sean, his self-serving welcome for help has worn thin because he is never willing to return the favour for others.

Sean puts his name forward in every municipal election year to run as councillor for his division. If someone else comes forward to run against him he withdraws. However lots of times due to public apathy and indifference nobody comes forward to honestly meet the needed challenge. As a result Sean is often acclaimed to be the councillor. His first time in office he made sure he got a brand new gravel road right up to his door, complete with an oiled patch up his driveway. Another time he manipulated the use of the RM scraper and operator for free, to trench some water from slough to slough in order to divert water flooding some of his land on to the neighbours place, far away from the natural flow of the land. In addition he convinced council to allow a dam to be built making sure much of the heavy spring drain-off went onto the neighbour's property. Sean enjoyed the paid trips to Saskatoon or Regina for the annual Saskatchewan Association of Rural Municipalities (SARM) convention just before Christmas. His wife could come along to shop and he would miss most of the meetings, between shopping and drinking up in the various hospitality rooms.

There is no doubt who Sean will be voting for in the next election regardless of the party platforms because, as he proudly says himself, "I am hooked into the Conservative party database and I get an e-mail or "robocall" every week on Thursday. I receive a list of talking points that keep me well informed about what is going on." Deep down, even Sean admits that robo-calls, directing people to the wrong voting station is a criminal offence, but he dismisses it as the new way of doing politics. Besides Sean, is happy with the board appointments and travel perks he receives when his party is in power.

Windy Bill, Gene the Genius, Len the Lefty, and Sean the Shyster almost always have a place at the front table in the White Bear Pub. They make for strange bedfellows, but as someone once pointed out, "We can pick our friends, but we can't choose our family or our neighbours." I guess we can't choose who goes to the pub either.

Over the years on my numerous trips to that prairie oasis, called the White Bear Hotel I have probably heard thousands of stories. Some of the tales have been told over and over again and have become valuable farm folk-lore. Others I just find interesting enough to remember, and I chuckle a bit when I get to tell them. I would love to share a few of those "I Remember When Stories" with you.

I remember when neighbours have learned to hate neighbours after liquor started to talk in that pub. Larry has hated Martin for many years now because Allan told Larry that his old friend Martin had seen him with his fancy red car stopped on the edge of the road with the hood up. He said Martin told him that Larry might as well have the damn trunk open instead of the hood, for all he knows about mechanics. Martin had said nothing of the sort, but Larry has never talked to Martin since. Whenever Martin's name comes up even twenty-five years later, Larry will say "That Martin is a lying f***ing bastard." Martin was not the liar, because it was Allen.

I remember when one of the new "store girls" arrived in Lacadena and one of the guys asked, "Did any of you fellows see the boobs on that new store girl in Lacadena? How would you like to be the guy who puts milk into those? No doubt she will be really good for attracting business. I guess she lives in that small travel trailer out behind the general store." Those store girls got teased a lot and all the single guys wanted to get to know them. Some of the girls liked it and encouraged it, while others didn't. The trick was to politely find out which way each girl leaned. Usually they were good sports.

When the last store girl that I recall lived there, I remember on Halloween night, we attempted to borrow a young Hereford bull from Fred's feedlot next to town and hook it to her little mobile home. However, the joke was on us once we got the rope on that larger than life bull and opened the gate. Three of us couldn't hold that oversized red bull with the white face, so he dragged us until my hands bled and my pants were nearly completely ripped off. He got away, but at least he was still in the larger fenced pasture. We don't know who got that rope off of him, and we were afraid to ask anybody, especially Fred. However, I can remember silently and carefully sneaking around Fred's house as we herded that stubborn bull back into the corral. So much for our trick on the "store-girl," the joke was on us, and my bruised legs hurt for days.

I remember when my friend Doug the plumber told me about an interesting experience he had the Thursday before a big Saturday wedding in town. I guess Joan from near Lacadena was cleaning her teeth and gargling to freshen her breath, when she accidentally spit her false teeth into the toilet just as she was flushing. Down they went, leaving her without top teeth, and, what was worse, she was needed to sing at a community wedding on Saturday. In a panic, she called Doug as if he was the 911 dispatcher. Doug told her "Don't worry my dear, I will be right there." He said he knew before he got off the phone just what he had to do. He hopped into his two-ton septic tank truck, which was painted yellow—the one we called the "Stool Bus"—and headed down the road. Joan and Jim were waiting for him when he got to their farm. "Joan was still awfully pretty as usual" Doug said, "but she was also very toothless," he added.

Doug quickly started his pumping system and began to empty out their septic tank. When the tank on the truck had pumped out Joan's entire septic tank, Doug told Jim and Joan to follow him to the grid road right by their yard. After he managed to get his heavy load up onto the road, he then pulled out his dumping hose onto the flat track behind the "Stool Bus" and advised Joan and Jim to keep their eyes on that hose. Doug slowly drove the tank truck forward, unloading the tank as he moved down the road. All of a sudden, a set of dirty false teeth came out onto the track on the road, as gently as a morning snowfall. Joan grabbed her valuable treasure with the fancy rubber gloves she had on and headed for the house. Doug told me with a twinkle in his mischievous eyes that Joan scalded and scrubbed those valuable chops, put them back in her mouth and people at the wedding say they never heard her sing more beautifully. Doug was quite a hero with the ladies after that. I remember when he told the bank manager's wife in Lacadena to please chew up her corn a little more because it was plugging up his pumper-truck. He advised her that the "Stool Bus" nearly choked to death last time he was at her house.

I remember hearing about swimming at the river and learning how those folks from the "Clytha District" did some crazy things. Folklore has it that when the young ladies were in the mood to choose a man, they participated in a courageous and revealing activity. The event would take place down at the river in the fast flowing channel. The evaluating women would sit right next to the channel with their old-style binoculars in hand. The young men would walk upstream about two hundred yards away and in the truly natural buff, float on their backs past the observing girls with their erect flags flying, if you know what I mean. That is what they meant by "ladies' choice" in the Clytha District.

I remember when everybody got those two-way radio sets as a means to visit and gossip from farm to farm. The set-up was like a multi-unit party line for anyone who dialed into the same frequency, and was brave enough to speak the truckers' language. "Do you copy, Whisky Jack? This is Tumble Weed." "Ten-four, Whisky Jack—let's go to channel seven."

"Roger." Everybody would follow them to channel seven to listen in on their brilliant discussion. To be fluent in that world became a real status symbol. I specifically remember when Clarence Smith asked old George from down at the ranch how his new farm hand was working out. For all the two-way audience to hear, George replied, "By geeze I can't say too much, the dirty son of a bitch is sitting right here beside me." I always wondered what else could George have said?

I remember many times hearing the story about that old wooden shed out behind the International Machine Dealership in Lacadena. Inside that antique shed there was a four-foot by eight-foot solid cement block. It was once used to anchor the batteries and charger assembly for the old windmill on top of the shed. Anyway, some of the wise guys in town decided they should get that gigantic hunk of cement out of that wood frame building, so they could use it for something else. "That's easy," said George McDonald, who was just back from the war and thought he knew a thing or two about dynamite. "Let me see what an expert can do for you guys," he bragged. "I can crush that cement so small it will come out in a kid's wagon ready for his sand pile." It sounded like a good plan. After all, George had survived four years of war at the front lines. So with the complete approval of the boys, as the crowd gathered, he got busy on Friday morning to rid that shed of the cement block. He drilled some holes, packed some mud, and inserted several sticks of dynamite with the caps in place. Everybody moved back and took cover. Finally, there was a blast like nobody there had ever heard. The street vibrated, glass dishes came off shelves inside houses and the sturdy shed flew almost two miles out of town. However, the cement block remains to this day as big and as solid as the day it was made, much to "Dynamite George's" displeasure and his diminished reputation.

Perhaps the most difficult decision you will ever make at the White Bear Hotel is deciding when it's time to go home. Like most pubs, it has a way of causing you to lose track of time. There are no clocks on the walls. However, at some point the traditional blinking of the lights will advise you to drink up and head home.

Heading back to Tyner, you go west down the scenic White Bear Hill where in daylight you can see in the southwest the natural beauty of the river flowing through its green tree clad valley. If your trip is after dark, you can see the reflection off the river and several lights dispersed around the flat plateau below, some from yard lights defining each farm, some from the nearby towns, or in the fall, some from the harvest machinery working late into the night on the vast fields. More recently, you can see the lights from compressor stations, sending natural gas from multiple well sites through a network of pipelines to market.

The Auction Sale

Farmers and town folks together came from miles around to experience the partly morbid and partly festive farm liquidation auction sale. The morbidity factor came from the brutal fact that one more farm family would disappear from the shrinking farm numbers. On the festive side the sale was a joyous meeting place for friends and neighbours to set aside day-by-day chores for a few hours of relaxation and informal socialization.

Parking spots would be limited in the farmyard itself, with all the cars, trucks, and trucks with trailers that arrived from all directions. Vehicles of all kinds with passengers of all ages, sizes, and shapes would park on the approaching road and into the ditches. These people may have seen the sale advertisement in the "Western Producer"—the farmer's newspaper/bible—or on colourful posters on every bulletin board for a hundred miles around. The sale would usually be scheduled to begin at ten am starting with the complete assortment of household belongings.

The "Auctioneer" and his work crew would have been busy for a couple of days before, organizing all the household furniture and appliances, dividing all the shop tools, hardware and materials into neat piles or pails, and lining up all the farm machinery from smallest to largest into a huge machinery row. The auctioneer would follow that exact same sequence as he sold one farm item after another. However, first in his well rehearsed preamble, he would name the farm family one person at a time and explain how long they had farmed and where they were going. Then he would say how well the farm property had been cared for and how it still had a long useful life.

In order to prepare for the bidding process, you would go to the auctioneer's trailer and sign up, giving your contact information in return for a bidding card with your number on it. That is the card you hold up to identify your bid as the auctioneer begs for your money. The local farm ladies group, also raising some money for their charities, would be set up in a nearby shed with the big door wide open, for sale patrons to stop by for a coffee, or juice, or pop and a slice of cake or pie. They would be in out of the heat or shaded from the cold wind and have several tables and chairs set up for visitors.

The auctioneer had a truck revamped like the "Popemobile," with a comfortable perch in the back for him to ride along the rows of sale items. His helpers, who made sure every motor would start, every tool would work and everything was clean, would now become sales helpers during the bidding process. They would yell back to him as bidders lifted their bidding card to signify a bid. As the sale got underway, the skilful auctioneer, in a loud rhythmic voice, would begin: "Who will give me fifty-fifty-yup—now sixty-sixty-yup—now seventy-seventy-seventy—SOLD—Sold for sixty dollars." Then it was time for the buyer to go over to the trailer where he or she got the bidding card in order to pay for the purchase. Every knife and spoon, bed and dresser, pail of nails, skill-saw and drill, fuel tank and welder, truck, tractor, cultivator, sprayer, swather, and combine were sold. The long sale day is over and another farm family is gone from their life on the land. It was like a going away party for a family and a funeral for a farm.

Sharon, Brandon, Danton and Dan Danielson in their new farmyard
(photo courtesy of The Saskatoon StarPhoenix)

four

GROWING OUR FAMILY FARM

Over the years I have sometimes told our own boys, "If you see a turtle sitting on top of a post, you know that he didn't get up there all by himself. He needed to have help." The same can be said about starting farming. The only question which needs to be answered is: "How much help and from where?" Some farmers have bragged that all they started with was a sack of wrenches, others say they knew just when to enter the industry and buy the land and machinery they required. However, generally speaking, all farmers got a start with various degrees of help from family, neighbours, banks or government assistance programs.

For some farmers, starting farming meant staying home, or coming home to a ready made turn-key farm operation with the best of everything available at the time. Remember that farming begins with a necessary land base, but it also requires farm machinery of many kinds, farm buildings of many kinds and a suitable place of residence. Not to mention all the small tools and short-line equipment needed to support the farm operation. In addition, a farm needs a self-contained water and sewer system to support the farm home and other farm buildings. This water supply may be from a natural well on the property or it might mean hauling water by truck and tank from a community spring, home to a large five to ten thousand gallon concrete storage cistern. Community water lines from a treatment plant at the river have now been installed to most of the farms.

Then there are the less obvious requirements such as a line of credit arrangement at the local Credit Union or Bank, as cash flow to cover input expenses each year before a crop is grown and sold. Finally there is the labour and management necessary to make things happen at the right time and place. There is crop and seed selection, developing a suitable rotation of crops and summer fallow, coping with the challenges of severe weather conditions, and storing or marketing the crop once harvested. The work at least at the start-up stage is almost always shared with other family members including the farmer's wife or partner and parents or other relatives. Often this in-family assistance is an unpaid contribution to the farm start-up initiative.

Finding A Land Base

In our case I am happy to admit that we got assistance from many sources. We had help from the federal Farm Credit Corporation (FCC) as a farm lender, the Saskatchewan Land Bank Commission, a farm land rental agency, the (PFRA) Prairie Farm Rehabilitation Agency a federal farm assistance program for water, sewer and shelter belts, as well as the Sask. Crown Corporations like SaskPower and SaskTel. In addition, several local contractors were hired along the way to keep our operation modern and growing. Even the local Rural Municipality got involved to build us a new farm access road. I suppose the real unsung heroes of this building process were the folks who actually agreed to sell us or rent us the basic land resource, without which no farm could exist. In our situation over the years something like a dozen land purchases or rental arrangements needed to be struck. Land was difficult to find and the price was always above productive value because the successful bidders in the market process often had extra income from outside the farming sector. My approach to purchasing farmland in the market place was based on the premise that in order to get the land, you must offer a high enough price that for two years the neighbours think you are crazy. Then, after two years, the same neighbours will say how lucky or smart you are for getting that land for that price. They might even say you stole it. On a couple of occasions I have not taken my own advice on that approach in the land market, and I have regretted it ever since.

The very first piece of the puzzle in the eventual footprint of our farm started with one-half section of land I cash rented from Leo Renner from Eston. My annual rent was $2800 dollars for 320 acres. The next piece in our puzzle was the purchase of the other half of that same section twenty-five, from Dick and Edna Sommerville for one hundred and ten dollars per acre. My parents were able to put up a half section of their land as security for the Farm Credit Corporation to lend me the money I needed. Then together with my uncle I bought the adjacent quarter section, the southwest quarter of section thirty-one. At that point I was farming four and one-half quarters of quite good farmland and also working at my city employment.

Real life would then provide some serious obstacles. Leo Renner suffered a heart event, so when reorganizing his finances, he and his wife decided they wanted to sell the land I rented from them. They would have been happy to sell it to me but I didn't yet have enough security to qualify for the funding. I was worried that I would be unable to continue to farm without that land base. However, consistent with the objectives of the new Saskatchewan Land Bank Program intended to facilitate the intergenerational transfer of farmland, Land Bank was able to purchase the

Renner Land and rent it to me. That was a pivotal event and lifesaver for my farming ambitions.

At roughly the same time, my uncle decided he didn't want to continue to share the work on the quarter we had purchased together, so we sold to the Land Bank, and I was able to also rent it as a family transfer. Then I was farming five quarters.

Another major piece to our puzzle came into place when my parents, now working in Saskatoon decided to rent three quarters of their land to me and Sharon my new wife, on a crop share arrangement. In addition, we rented one-half of another quarter from them that my dad and I had just purchased together from John Case of Outlook. Just before that I had bought the south-east quarter of section nineteen from Glen and Doris Blood. It would soon become our new farmyard. Not long after the establishment of our yard a neighbour from Lacadena, George Case the brother of John Case, advertised for sale, the northeast quarter of section nineteen which was right next to our yard. On his advice we bid seventy-five thousand dollars and got it. That deal increased our farm to all of section nineteen, all of section twenty-five, half of section one and the southwest quarter of section thirty-one, for a total of eleven quarters.

A few years later, Lee Britton, who farmed beside us, called me about selling us the south half of section one as the last piece of our thirteen-quarter 2080-acre farm puzzle. Incidentally, with the increased value of farmland, and our farm growth, the FCC released the land back to my parents that they had put up as security for my original Sommerville purchase. Sometime later after the passing of my parents, Sharon and I inherited one more quarter section that we were already renting. Two other quarters went to my brother and my sister that we continue to rent from them. Another unexpected event took place a few years ago when ten gas wells were drilled on our farmland. From them we receive an annual lease income of about twenty-five hundred dollars per well.

One of ten gas wells

Building a New Farmyard

Prior to the purchase of the land for our new farmyard, I was a resident of Tyner living in a nice little home I had purchased a few years before, for seven thousand dollars. Then a very attractive and accomplished banker from Saskatoon would come into my life. Sharon Oliver had grown up at Greenan about thirty miles from my hometown but I didn't meet up with her until she had gone to work for the CIBC in Perth Ontario and then transferred to the bank in Saskatoon. While at one of my off-farm winter jobs in Saskatoon, we met in the bar at the Senator Hotel. To make a long story short we teamed up to improve my farm and grow our happy farm family, with roots down both at the farm and in Saskatoon.

My mom and dad were very happy about this new team effort. They had catered to my bachelor uncle for most of his life, and felt that the necessary motivation and ambition for productive farm life, was not generated by the lonely existence of a farm bachelor. My dad in particular held that view and became more in favour of me continuing my farming career once Sharon and I were married. He was one of those guys who worried that if you lived alone and woke up with a bad idea you might still have the same bad idea when you went to bed at night. My mom was overjoyed with my choice of Sharon as my life's partner.

My parents had always farmed with my uncle, with whom they could share resources for the land base, machinery and labour. However, it became obvious to me that my uncle did not have the ambition or interest needed to expand and sustain a farm that would include us all, including Sharon and me. More than a few times I had seen opportunities missed because my uncle played the role of a contrarian like an anchor holding my dad and mom back. In some ways my uncle was more like another kid for my parents to look after rather than a productive business partner. When I attempted to join my dad and my uncle with a desire to expand the farm base I came up against the same kind of resistance my parents had always experienced from my uncle. However, I was too young and aggressive to cope for very long with any negative influence. Therefore, I soon asked my dad and mom to decide if they wanted to continue to farm with my uncle without Sharon and me, or if they would like to partner with us. I don't think it was an easy decision for my dad after so many years but it was easy for my mom. They quickly decided to come with Sharon and me. We were extremely glad and very fortunate that they made that choice. My dad and I worked very well together and Sharon and my mom got along like mother and daughter.

In 1984 my dad reached age sixty-five and retired from his job at the University in Saskatoon, just in time to return to his love of farming with Sharon and me, as we continued to grow our farm. The arrival of our boys, Danton in 1982, and Brandon in 1983, provided an additional incentive for grandma and grandpa to be with us at the farm for the busy summer months. Grandma would spoil us with delicious baking, cooking and cleaning, while grandpa contributed as he stuck to that tractor seat like glue.

As good farm luck would have it, right about then a primitive little two-bedroom house, owned by the Sask. Wheat Pool in Tyner came up for bids. We placed a "stink bid"—as they are sometimes called-for three hundred and fifty dollars. Since it was the only bid submitted, we got the house and set out to renovate it into a summer cottage for my parents. My mother would let her homemaker talents shine as curtains went up, walls got painted, and windows were cleaned. For our part we provided second hand but good appliances, kitchen supplies, the telephone hook up, and a new shower assembly, since the house never had one before. This old but new home concept clearly made my parents happier for their retirement years. They became a tremendous asset to the success of our family farm.

Starting a farm on a new farmsite as we did offers many complicated challenges. Our new farmyard was actually an old homestead of a pioneer family who had planted trees and lived there for several years but vacated in the "Dirty Thirties," leaving behind only the shelterbelt, one cistern, and one large twenty by thirty-six foot grain bin. That bin, with some renovations, soon became our workshop.

The provincial crown corporation, SaskPower, installed the poles, wires, and transformers to get the electricity to our new farmsite. Most of the line was assembled underground arriving at only one pole in the middle of our yard. That yard pole became the anchor for the main yardlight, which is switched on at dark by an automatic electric eye. The provincial telephone network was connected to our new yard by Sasktel, another provincial Crown Corporation. The federal government Prairie Farm Rehabilitation Agency was also there to assist with the cost and delivery of plumbing equipment like pumps, tanks and pipe for our water and sewer systems. In addition, PFRA provided assistance with the cost of digging dugout storage capacity for farm water, and also provided many trees to enhance our farm shelter belts. The local Rural Municipality of Lacadena #228 had the responsibility to build and maintain a farm access gravel road to our farm site. The establishment of new farms and new farmers at that time was seen as important, because we were considered an enhancement to our economy. Large government investments were viewed as good for the future prosperity of the community, the province, and the country.

The first temporary house in our yard would be a fourteen-foot wide by seventy-two-foot long, new furnished mobile home bought in Kindersley.

Since there would be no basement under the mobile home it was necessary to dig in a four-foot diameter well casing to hold the pressure system, sewer pump and pipes, below the frost line. In addition, the power lines and the telephone lines inside our yard would be installed underground. We built our forty-eight-foot wide by seventy-two-foot long, wood quonset as a machinery shed in 1982. Over time, we purchased about ten large steel bins, some with hopper bottoms and some with flat wood floors. In 1986 Sharon spotted an old farm house over near Isham about twelve miles west and north from our place, which she decided would make a great farm home for our new yard. She felt it would be a good permanent replacement for our mobile home, which she agreed had served us well for six years. I had previously sold my little house in Tyner to buy that mobile home for the farm so it was paid off. We could sell it for sixteen thousand dollars as a financial start on our new to us house-moving project. That would be a forty thousand dollar project.

Moving our new farmhouse

First we bought the house for four-thousand dollars and paid about nine thousand to have a new full concrete basement dug and completed. Then a local mover, Arnold Petka from Beechy, agreed to move the house for five thousand-two hundred dollars, if I organized a work crew to help him and his young assistant. That was indeed a labour of love and a valuable learning experience for Sharon and I. It came down to hard physical work blocking and bridging underneath four powerful hydraulic jacks, capable of lifting the weight of a whole house in unison with each other. Then the large steel "timbers" as he called them would be slid under the now lifted building, and the house on timbers would slide onto the trailer wheels attached to the powerful moving truck. Within three or four days we had the power, telephone, water, sewer and everything else unhooked at the old location and the house was on the truck heading down the road to our new farmyard. One power line needed to be raised but within seven or eight hours it was over at our place. Once over at our yard, that same previous house loading process had to be reversed, as we slid the house off the truck-trailer assembly onto the new basement. The whole unhook, move and re-hook procedure lasted a long week.

However at that point the job seemed like it had just begun, because it was time to replace some old windows and doors, put in a new furnace, construct a new bathroom and do some painting. As I said it was a labour of love with lots of family and neighbour support. We were young, ambitious and completely naive about such things. I am now certain that such a job is not for those of us getting a little long in the tooth. However, at that stage in our lives it became a rewarding experience that Sharon and I will never forget. Unfortunately about that same time we lost Sharon's mom, who faced a difficult battle with cancer. However, as part of her mom's legacy, Sharon received approximately twelve-thousand dollars from the estate, which contributed a good share to the relocation and renovation of our new farm home.

All together this house relocation in 1986 cost us right on forty thousand dollars. Sixteen thousand came from the sale of our six year old mobile home, twelve thousand came from Sharon's mom Sally Oliver and ten thousand came from a provincial government home renovation grant. The other two thousand came from our limited savings.

It is with great respect that I thank Doug McMillan the local plumber, handyman and septic truck man for devoting almost his entire summer of 1986 helping Sharon, our boys and me acquire our new farm home. Our existing mobile home had to be unhooked, moved out of the way, and re-hooked to the water and sewer for our use while the house moving project was underway. Then the "new old house" at its old location near Isham needed to be unhooked from power, water and sewer, so it could be jacked up and loaded on the moving truck. After our "new old home" arrived there was no end to the plumbing, electrical and carpenter work that needed to be done. Doug loved his almost constant coffee breaks and lunch breaks, along with all the gossip we could attract. Somehow with the help of my dad, some friends, and Sharon and I, he got everything done that he bravely tackled. His help was greatly appreciated.

Doug had a son-in-law who turned up from Newfoundland. Amus or "Famous Amus" as we liked to call him was a good natured, hard working fellow, who was attempting to deal with an alcohol issue. Doug asked me if we would give him a job while we completed this house-moving project. Taking Amus on was probably one of the best spontaneous business moves we ever made. He could do anything from hard labour to precise carpenter work. Amus painted the roof, put in new windows, hung a few new doors, replaced damaged siding, helped with the plumbing and just helped out with everything. He could hang from any "skyhook" and manipulate a new upstairs window into place. We liked each other and had fun getting big jobs done. What I remember most is the day after the main house was moved when Amus and I set out with a fragile, shaky, borrowed bale wagon to retrieve the porch which had been left back at the old location. Two dangerously undependable "Jackalls" were all the technology we had with us. Moving that porch with the backside open ready to be reattached to the house was a herculean task even for famous Amus and me. During that unceremonious move was the first time I ever heard a "Newfy" swear his guts out. Most of the sailors from his native land on the "Rock" would have been proud, and most prairie mechanics would have been speechless.

This mammoth house relocation task ending in our new farmyard as our new home was not the only time this old house had been moved. This newly renovated home for us had once been the general store in the now gone town of Tuberose, about twenty-five miles east and south of our place, near Kyle. From there it had been moved to the Martin Farm as a farm home about fifteen miles straight east of our farm on the other side of Lacadena. Then in the early sixties, it was moved again to the Sheasby farm by Isham, about twelve miles north west of where it now rests in our yard.

It would be easy to become envious of other farmers who walked into a turn-key farm operation, complete with house, land, farmyard, machinery and tools. There was a lot to organize, plan and purchase when it came to setting up a modern farm from scratch. Nonetheless Sharon and I along with Danton and Brandon are very proud of what we were able to accomplish, with hard work and off-farm income. We are also very appreciative for the help we got from my parents-Urvan and Geneva Danielson.

The Farm Credit Corporation

In circumstances where a farmer is in a financial position to provide already-owned farmland as security for a loan, purchasing farmland has been a good investment over time. Most expanding farmers turn to the Farm Credit Corporation (FCC) to fund those purchases. In my case a half section of land put up by my parents as security allowed me to purchase my first land through the FCC.

The FCC is a federal government crown corporation, which essentially operates as a bank for Canadian farmers. It will take a chance with farmers who otherwise may not be able to secure financing for farm expansion from the regular credit union or banking network. A major benefit from FCC is its ability to lend up to or very near to the full price of the farmland being purchased, thereby reducing or eliminating the required cash down payment. Interest charges are usually more favourable with FCC. In addition FCC offers financial guidance and provides a network of advisors who understand agriculture and keep up to date on world market conditions, the overall Canadian economy, and local land prices. The mandate for the corporation includes the recognition of how important secure food production and the success of the farm sector are to the sustainability of the Canadian economy. Much of our farm production is exported to world markets, which attracts foreign currency into our country, maintaining a positive export/import balance of payments and a stable Canadian dollar.

During very tough times a few years back FCC operated a Lease Back Program. Farmers could turn over ownership of their land to FCC and lease it back. In most cases a FCC Lease Back Agreement was a short-term workable solution for the farmer's financial problems. The good news is that often after paying a fair rent during the Lease Back Program, the farmer was able to negotiate the repurchase of the land under better FCC payment arrangements than had been the case originally. I must confess that our farm was served well by FCC and its lease back program.

The Saskatchewan Land Bank

In circumstances when a farmer doesn't have enough equity in farmland to outright purchase more land, the addition of rented land becomes a realistic option. In our case, we were fortunate enough to be setting up our farm when the Saskatchewan Land Bank was in existence during the 1970's.

As I recall, the Land Bank idea originated at meetings of the National Farmers Union, The Saskatchewan Wheat Pool, Co-operative Implements Ltd. and the New Democratic Party. Then it was officially adopted as one plank in the 1971 provincial election platform of the NDP known as "The New Deal for People." The NDP were elected, and with the leadership of Premier Allen Blakeney, the Land Bank Program was implemented in 1972.

The average age of Saskatchewan farmers at the time was around fifty-five years of age. Many farmers wanted to retire and turn the farm over to the next generation. However, the problem was that if the older farmers gave the land directly to the younger farmers, they would not have the fruits of their life-long labours on which to retire. At the same time the new generation often did not have the resources or the collateral to acquire the funds to purchase the land even at a fair price. The Land Bank provided the missing link because it could buy the land at a fair market price from the retiring farmers, then lease it to the younger farmer on a long term lease, with the option to purchase the land at fair market value at any time during the lease.

Fair market value was determined by the average price, of the three most recent sales of similar land as close as possible to the subject land.

The Land Bank invested approximately forty million dollars per year for ten years, to purchase about 1.2 million acres of farmland, from twenty-five hundred vendors, to be leased out to over twenty-six hundred tenant family farms. Some of the tenants qualified as direct descendants of the vendor, being a son, daughter, grandson or granddaughter. In situations where there was not a direct transfer of a lease to a descendant, the lease would be allocated by a public competition. It was quite common for a dozen or so applicants to compete for one lease. A full double-page advertisement for "Land Bank Land" up for lease by competition would be published in "The Western Producer," a popular weekly farm magazine. Applicants obtained assistance with their application from Land Bank Councillors located at regional offices established across the province. Once complete the applications would be evaluated and scored by the Land Bank Commission relative to well prepared objective criteria, described as "the point system" in order to provide fairness and avoid claims of political interference. There was also an independent "Lands Appeal Board" made up of other farmers to ensure that the objective criteria was accurately applied.

The rationale for the competition criteria was to make sure the tenant would have enough land base to survive and grow his farm, but would not be set up with more land than was thought necessary. In some cases the land to be leased by competition would be divided to assist the greatest number of new farmers. Evaluation scales were employed to compare other components such as access to machinery, education and technical skills, tenant age, distance to other land, and whether the potential tenant had already been leasing the land from the seller to the Land Bank. In addition, a potential tenant could receive some points, exclusive of the other applicants, if the vendor designated that applicant as his or her preferred applicant. The rationale for such criteria was to minimize any disruption to existing farms, if a landlord found it necessary to sell.

The original leases began as a cash rent arrangement set at five percent of the price the Land Bank had paid for the land. However, it was soon changed to a productive rental formula which was paid as a cash rent. Together with cash rent and taxes, the payment came very close to the traditional one third of crop sales. That tradition was based on the age old concept that one third of the returns from farm production should go to the land owner, while one third should go to the farm operator for input costs and the other third should go to the operator as income for labour and management. The reason for the change from the percentage of land value rent calculation to the productive rental formula was to relate rent costs to actual returns from production, not inflated land values. The buildings on the Land Bank land were sold to the lessee at the depreciated value over a reasonable amortization period. That turned the decision making for upkeep or removal of buildings over to the tenant. If the lease was vacated, the tenant would be compensated for the depreciated value of any improvements.

Nobody I knew has ever argued that having rented land from the Land Bank was more beneficial than owning land outright, without rent to pay and no debt owing. However, that ideal ownership objective is not always available to new young farmers. The Saskatchewan Land Bank was a pragmatic solution to assist both retiring farmers and those who wanted to enter the farming business. For many retiring farmers this program was a dream come true. It was an opportunity for them to retire with dignity and move to a new life in town or the city receiving a fair price from their farm. The bonus feature for them was also having a family member or another young family take over their family farm, entering a career they also loved. For many Saskatchewan farmers like ourselves it was the Land Bank which provided our start in farming and set the foundation for the accumulation of other land.

The main pillars of the Saskatchewan Land Bank Intergenerational Land Transfer Program from my point of view as a tenant were:

1. long term lease security with the right of transfer to descendents
2. fair productive rental formula (a traditional 1/3:2/3 share)
3. a realistic option to purchase
4. independent decision-making on farm management
5. recognition of the leases as security at credit institutions.

The closer the Land Bank could come to emulating the benefits of outright land ownership, the more the tenants would care for the land as if it was their own. It was a long-term agreement with benefits for both the government and the farmer, so the agreement would be sustainable.

Farmers of all political affiliations benefited from the Land Bank an idea whose time had come. The biggest obstacle to the success of the program centred on whether the farm population and the general provincial population understood the positive impact it could have, or if they saw it as a political football to be kicked around and destroyed. There were many people who opposed the program. For some it was personal sour grapes, as twelve or so potential tenants would compete for a lease, but only one could be successful. That numbers game produced animosity toward the plan and the government that implemented it. There were some people who said the Land Bank would own all of the land in the province, even though only two percent of the farmland was purchased in a decade. At that rate it would have taken five hundred years to purchase all the land, as some had feared, even if every farmer in the province voluntarily chose to sell, and infinite funding was available. That ridiculous scenario would also need to assume that no tenants would buy out their leases with the option to purchase that they were guaranteed. Not very realistic, based on common sense. However, unfortunately common sense is one of God's gifts that he/she does not give out equally to all people.

Others also opposed the Land Bank for even more questionable reasons. There were "land hogs" with a "cannibal nature" wanting to profit by devouring their neighbours' farms at fire sale prices; especially if that seller was in a desperate situation, due to finance or health issues. There were private sales agents, who profit from regular farm sales. There were land transfer lawyers who thought they would lose business. In addition, there were some ambitious politicians, financed by some of the previously mentioned profiteers, who set aside reason, for what they hoped would be political gain. None of these negative influences were consistent with the farmers' best interests, especially the best interests of retiring farmers who wanted a fair price for their land.

Some politicians and others egged on by the politicians claimed that the Land Bank pushed land prices out of reach for other farmers, even though by design the purchase price was always the average of past sales comparables. As a result, prices the Land Bank offered followed rather than led the market. In reality it would eliminate fire sale, lower than reasonable prices, forced on those who needed to sell in an emergency. Some politicians even played the fear card driven by political dogma, suggesting that the Land Bank was a communist plot, returning us to the plight of Eastern Europe from where their ancestors had escaped. Yet another criticism of the Land Bank from politicians and others was the notion that the government would be telling the farmer what crops to grow and how to grow them. Nothing could have been further from the truth.

Reasonable people given correct information knew what part the Land Bank played in saving the family farm. They knew all farmers feel that outright ownership of farmland is the ideal and is a highly desirable dream, but in the real world, lease tenure coupled with private ownership is a fact of life. Long-term leases with security of tenure until age sixty-five, with an option to purchase, were a highly desirable alternative. That is why so many farmers applied for any leases that came up in a competition. Most farmers rent some part of their farm base, and are always looking to lease more. However for the reasons mentioned the Land Bank became a highly charged "political football." Over time the negative forces came together to cause the demise of the Land Bank experiment.

The opposition party of the day could not stand the success this program enjoyed. Grant Devine, as leader of the Saskatchewan opposition Conservative Party, made this program a major political issue in the 1982 provincial election. Tenant farmers and sellers to the Land Bank saw their lives, without justification, dragged through the political dirt. The Conservatives kept repeating what Land Bank observers knew was not true and completely irrelevant to the situation. Devine kept saying it was better to own land then to rent it. The real question was, if you can't own it initially, isn't it better to rent the land with long-term security of tenure and then buy it when you are able to? After the election the Devine Government came to power with a lesson of scorn and revenge to teach the so-called socialists connected to the Land Bank. Almost immediately upon forming government, the Conservatives passed legislation to dissolve the Land Bank.

The Saskatchewan Crown Land Tenants Association

When the NDP government decided to change the Land Bank rental formula from a percent of inflating land values to a productive rental formula, "The Crown Land Tenants Association" formed to advise the government and present the views of the tenants. Over the years that association met from time to time to represent tenants. When the Devine Conservatives came to power in 1982 and then eliminated the Saskatchewan Land Bank, the Crown Land Tenants Association quickly came alive to help save the existing leases for its farmers.

In early December of 1982, the association organized buses, trucks, and cars filled with Land Bank Tenants and other friends of the Land Bank, to attend a large rally at the provincial legislature in Regina. It was an all day event beginning as a march at the Legislative Buildings and ending at the Centre of the Arts for an organizational meeting. It soon became obvious to everyone including the government, from the individuals present, and the heartfelt speeches, that many members and supporters of the provincial Conservative Party also had benefited from the Land Bank. That fact spoke volumes for the "point allocation system objective criteria" which provided Land Bank benefits based on the objectives of the program, rather than partisan politics. We can thank that fairness objective of the program for many Conservative supporters who challenged their own governing party. Thanks to these farmers and friends the existing leases were protected.

On a personal note, that December 14, 1982 date of the Land Bank Tenants Protest in Regina was extremely significant for our budding family. As one of the organizers of the event I felt I must attend even though Sharon was nearing the delivery date for our first child. We had already spent one night in the Eston Hospital a few nights before due to a false alarm delivery. I remember sleeping on the operating table, which was the only available extra bed not in use that night. The Hospital by then did not do operations anymore but it would deliver babies. Without any delivery success that night we were offered a house by a friend to stay at in town to await our little bundle of joy. However, I was called off to Regina.

Later in the day we had all moved over to the Centre Of the Arts to continue the meetings. Someone passed a faintly scribbled note up to me at the front of the Hall. It simply but emphatically said "Tell Dan Danielson that his wife is on her way by ambulance to the University Hospital in Saskatoon to have her baby." Since I had travelled from Kyle to Regina in one of many chartered buses we had arranged for the day, I didn't have ready transportation on my own to go anywhere. Wayne Gamble from Perdue was near me when I read out the note. He quickly said that he had his truck there and would take me to Saskatoon. Within seconds I guess we said our hasty good-bye and off we went. Thanks to Wayne, a great friend, I arrived at the hospital just as our important human delivery was already taking place.

In those days it had not yet become routine for the dads to be in the delivery room for the event but some doctors encouraged it. As a result I had been shown to a soft chair in the waiting room next to the delivery room. Right then as my mind swirled about the Rally and I contemplated becoming a new dad, the doctor came in the room. He looked like he had just committed a gruesome murder and forgot to destroy the evidence. He said: "Mr. Danielson you have a big, beautiful, healthy son in that other room, and Sharon is doing well." Then he said: "I have some finishing up to do in there but you can come in with me if you like." When he escorted me into the delivery room I saw Sharon's lovely smile as the nurse handed me Danton Lorne Danielson (9 pounds plus) tightly wrapped in a warm flannelette sheet with just his face sticking out. Once again like so many times before and afterwards, I was reminded how good life has been to me. A few weeks later as I told this story in the White Bear Pub; I bragged to the group that a good farmer always gets his seed back and more. I must not forget to say how proud Sharon and I were four days less than one year later when our other healthy son Brandon Darrell Danielson was born in Saskatoon.

There is no doubt in my mind that the Land Bank was generally an overwhelming success for farmers and communities across the province. Unfortunately its success in some ironic and confusing way was a major cause of its downfall. The desire of farmers to access the benefits of the plan outran the funding provided to satisfy the need. As land prices increased less land could be acquired by the established funding policy and the government was reluctant to increase the funding enough to be as effective as it once was. More senior farmers applied to sell than could be accommodated and more potential tenants applied than could be allocated leases. That success was at the heart of the criticism by the public, who saw the Land Bank as becoming ineffective. Unfortunately, the increased budget requirements to purchase higher-priced land and the increased public criticism, as illegitimate as it was, had caused the government to step back from the defence of its own successful experiment in intergenerational land transfer. The NDP were beginning to retreat from their once aggressive policies for the protection of the family farm. In response the Conservative party was prepared to use incorrect information prompted by right-wing ideology to defeat a useful program for political reasons alone.

Co-operatives And Our Family Farm

In the time period when Sharon and I established our family farm we very much respected the spirit of co-operation as it was demonstrated in organizations like retail co-ops, the Saskatchewan Wheat Pool and the Canadian Wheat Board. We appreciated what they had to offer us, and our communities. Both Sharon and I were proud of the work our parents and others had done before us to build and maintain the co-operative network. We took comfort in the fact that we could help control our farming destiny by getting involved to do our part in the democratic structures of the co-operatives, guided by some principles we supported.

The basic guiding principles of a co-operative, which Sharon and I still consider very valid are:

1. Co-operatives have voluntary and open membership, available to anyone able to use their services and comply with the responsibilities of members.
2. Co-operatives have democratic membership control with equal voting rights based on one member one vote in the decision-making processes.
3. Co-operatives have equal economic membership participation based on a set membership fee. Then, members are allocated a share of the profits after costs are paid and reasonable reserves are set aside, based on the proportion of patronage or business they have done

with the co-operative over the past year. That patronage allocation is divided between equity held in the co-operative and a cash pay-out called a dividend. When a member leaves the co-operative or reaches retirement age, the saved equity portion is paid out to the member, sometimes over a five-year period.

4. Co-operatives are locally owned, as autonomous, independent democratic organizations.
5. Co-operatives provide education, training and information to their members, their workers and the general public.
6. Co-operatives co-operate among co-operatives at the local, regional and national levels such as retail co-ops doing business with wholesale co-operatives or credit unions.
7. Co-operatives participate in sustainable community development.

Some of the co-operatives which my farm family have dealt with and personally relied on over the years were: the Lacadena fuel and hardware co-op, the Eston Co-op grocery store, drug store, gas station, farm fuel service, hardware, car and truck dealership and lumber yard, the Rosetown Co-op Implement Dealership, and the Kyle Pioneer Co-op fuel, hardware and grocery services, also affiliated with Swift Current Pioneer Co-op. We have dealt with the Eston Credit Union and the Saskatoon Credit Union. Up until the demise of the Saskatchewan Wheat Pool, all our grain was delivered to that company in Isham, Tyner, White Bear, Lacadena, Rosetown, Moose Jaw, Kindersley and Saskatoon. All of our grain sales went through the Canadian Wheat Board, a farmer owned single desk co-operative, functioning as a farmers' international and domestic grain marketing agency.

In hindsight, I submit that there are two very different pathways one might take on the journey to establish a successful family farm. The first option is a rather simple and smooth road, without much personal stress or cost. It is when a new farmer can take ownership and control of an existing farm operation, complete with land, buildings, machinery, and usually some additional family guidance and financial support.

The second option for farm development is the bumpy road Sharon and I were on for nearly forty years. It is the uncertain path toward farm success, which includes acquiring land, buildings, machinery, and then adding value over time, from almost nothing except planning, management, and hard work. Now we have farm land equity, our farm home and yard, our home in Saskatoon, our cottage at Christopher Lake, our FutureScope Management Consulting company and a self-directed pension plan to show for it all. We are not ashamed to acknowledge that in our case: "The turtle couldn't get up on the post all by himself." A reasonable amount of help was needed and appreciated as a good investment in our future. Outside assistance for farms in our day, was provided by our governments and our society, because of what farmers would contribute to the community, the province and the country.

There is a constant level of stress on farmers who confront the weather, the markets, the local politics, the new environmental realities, and the knowledge that co-operative and rural values are being eroded. However for me there are still many joyous memories that flash back to me from my days on the farm. I remember the freedom you have on the farm in the wide-open spaces. I remember the smell of freshly ploughed soil as you screen it through your fingers and think of how it was produced from previous crops over time. I remember driving the tractor pulling a discer, air-seeder, cultivator, rod-weeder, sprayer, swather or combine. I remember lining the tractor up with a distant land mark such as a building or tree to get a near perfectly straight first row of cultivated ground as the guide for the rest of the field. The neighbours will be watching and do comment on such things you know. I remember the quick but strong "whirl winds" a miniature tornado-like dust storm, that would almost instantly flash through our yard. I remember being upstairs in our farm house with the windows open during a thunderstorm, smelling the rain filled air and listening to the rain drops pound down on the roof. Then I remember watching a multi-coloured rainbow in the bright blue sky after that brisk thundershower.

I remember slipping and swerving down a muddy road manipulating our vehicle to stay on the high side of corners and curves. I remember unexpected visits from neighbours who just stop in for a coffee or drink after driving around in the evening looking at the crops. Maybe they were checking to see if there was any hail damage. I remember riding bikes as a family up the road and back for no reason other than the outdoor experience. I remember that huge orange Tom-cat that really never had a home but lived in all the yards for miles around. I remember hoeing in the garden after supper among the bees, butterflies, potato bugs, flies and mosquitoes to eliminate weeds from our generous crop of vegetables that always were more delicious straight from the garden. I remember the family conversations we had sitting together as we shelled vast containers of garden peas. I remember after being away for the day arriving back home late at night to be greeted by our four yard lights. I remember the deer with shiny reflective eyes grazing on the luscious lawn by our cozy farm house. I remember getting out of our car to the sounds of the multiple frogs croaking from the dugout and the choir of swans, geese, ducks or cranes that backed them up from the nearby slough. I remember settling into bed so proud of what I had in life. I remember when the yellow-breasted meadowlark, with the wide black V on his chest, sang his cheery morning song, from a post near our window. I still remember how beautiful farm life can be.

The Family Farm

by Dan Danielson

Farm
Good-Life
Your Own Boss
Exercise, Sunshine, Nature, Income
Freedom

Family
Our-Team
Excited, Ambitious, Committed
Seeking Security Of Tenure
Future

Land
Resource-Base
Purchase, Rent, Work
Investigate, Calculate, Communicate, Wait
Speculate

Machinery
Horse-Power
Expensive and Necessary
Bearing, Sprocket, Chain, Wheel
Deal

Finance
Net-Worth
Principal, Interest, Taxes
Collateral, Amortization, Payment, Equity,
Ownership

Government
Farm Support
Federal, Provincial, Municipal
Farm-Credit, Land Bank, Credit Union, Bank
Security

Town
Meeting-Place
Supplies, Markets, Services
Friendly, Healthy, Wealthy, Busy
Community

Enjoyment
Good-Friends
Visiting, Dancing, Hunting,
Family, Freedom, Nature, Neighbours
Happiness

five

FARM POLITICS AND THE FUTURE FAMILY FARM

Farm historians have often joked that farming is twenty percent food production and eighty percent politics. Political decisions resulting in policies that affect farm input costs or product prices or agronomic production practices are crucial to the success or failure of our farms. I believe the future success of our farming industry will depend on the willingness of farmers to come together with the courage to confront some harsh political realities to do what is best for farmers. If farmers rely on others such as the multinational monopoly corporations with their own profit motives to design our future, I worry that the results will not be positive. Those corporations, organized in monopoly structures and powered by the strength of their accumulated wealth and influence, will orchestrate a self-serving farm industry protected by one-sided international trade agreements, and sympathetic governments whose power they can purchase, with the assistance of high price lobbyists. The scenario they create will be reinforced by the best media spin money can buy. Ironically, they will convince a few vocal farmers and their rump organizations, that can't resist the tempting offers of special treatment and privilege, to betray their fellow farmers. Some folks are ready to accept wealth and power at the expense of their neighbours and their communities. Even if that promise of special treatment is merely a composition of meaningless rhetoric.

Historically, in response to large corporate monopolies and oligopolies established by the railroads, banks, grain companies and retail supply companies, some charismatic and courageous farm activists emerged to organize locally owned and democratically governed co-operatives, to provide direct farmer participation in the farm economy. As a result organizations like the Saskatchewan Wheat Pool as a grain handling company, and the Canadian Wheat Board a marketing monopoly were established. Local credit unions and retail co-operatives also came on the scene. Farmers had a window into the industry as well as their own competing enterprises. Profits from their co-operatives would come back from a farmer-organized orderly marketing system in the form of patronage dividends or final payments. That "Progressive Farm Movement" as it became known, had the benefit of supportive elected members of governments who encouraged this new farmer owned competition in the market environment. Those politicians felt comfortable with government involvement in the economy, so they cultivated a fertile environment for many rural development programs to assist farmers. Cooperatives and Crown Corporations together were seen as a symbol of market power, designed to help farmers compete in a harsh climate far from a distant world marketplace.

It was during my university days while working at the Collins Farm in the summers that I cut my teeth on active provincial politics. My boss and I discussed almost all the current political issues in some detail over many lunches and cold drinks. Large baloney, lettuce, and mustard sandwiches with ice tea, provided the nourishment. At that time my entry into the industry as a self-employed farmer entrepreneur seemed virtually impossible given the financial barriers to entry. However the evolution of the family farm was a frequent topic of our conversations. I should confess that by then I had adopted a critical if not negative attitude, toward politics and politicians, ignited by a cruel battle which I never quite understood, over the implementation of government paid medical care insurance. The Lloyd CCF/NDP government while implementing that universal and comprehensive program for the benefit of the poor and everyone else wanted the cooperation of doctors and other health care workers to make it happen. Their expectations were for a smooth transition. However, the process turned into a dirty fight with the doctors who were backed by the Liberal and Conservative political parties.

The Liberals and Conservatives had formed militant "Keep Our Doctors Committees (KODs)" and spread hatred against the government during that health policy crisis. It has been reported that the American Medical Association (AMA) sent millions of dollars up into the hands of the (KOD) Committees in a failed attempt to prevent the introduction of medicare. Ever since that senseless crisis I have been angry to think that our doctors and our Liberal and Conservative politicians would viciously fight to keep Saskatchewan people bound in an American style system of costly private medical insurance, only for those who could afford the high premiums. I guess what had soured me the most on this type of mindless politics was knowing that the health care of our people was being used as a wedge issue, driven by doctor self-interest and petty Liberal and Conservative politicians who tried to capitalize on that injustice.

In my opinion from that medical care crisis onward, the Liberals and Conservatives have continued an "Anybody but NDP" approach to Saskatchewan politics. It is a strange coalition of what I believe to be a dichotomy of principles that has confused and bothered me for many years. That strange amalgamation of Liberals and Conservatives in a right-wing partnership eventually resulted in the defeat of the CCF/NDP government by the Liberals under leader Ross Thatcher in 1964. In my opinion, ever since that public medical care insurance battle, a right-wing coalition has survived in one form or another up until today. Currently it has been reincarnated as the Saskatchewan Party.

Politics for that right-wing coalition, as I see it, is pandering to a few wealthy corporate masters, whose profits always come first before farmer welfare. There is always that notion that you should be able to buy yourself to the head of the line, and democracy should be privatized into the control of the better off. Profits are supreme even if it means pitting the economy against the environment. Co-operatives, crown corporations and government support programs for ordinary citizen welfare are viewed as obstacles to corporate profit and need not exist. However tax breaks and corporate welfare schemes are considered acceptable. Complete trust is placed in the hands of corporations, guided by a business dogma trusting that what is good for multinational monopoly corporations is also good for farmers. To the extent that public enterprise is tolerated, it is expected to help minimize corporate costs and risks, but privatize the profits, often outside the province or country.

Legislation and regulation protecting national sovereignty, consumer rights, financial equalization or the environment are not desirable in this "ideal business world" of deregulation, where corporations offer to police themselves. "Free enterprise" which is not actually free, and is becoming very expensive, if you ask the poor and middle class, is the media driven battle cry for this coalition of the right. Not much is said by those who have gained considerable wealth from our economy about the assistance they receive from our public schools, public health, public roads and infrastructure, lucrative tax breaks and public support programs. Yet, when asked to pay taxes back into the public purse, for public services, there is an unethical reluctance and harmful avoidance.

Over the years the political lines were drawn deep in the prairie snow, even defining which businesses and services should be patronized by which political factions in the rural communities. On the one side there were those who supported the Sask. Wheat Pool, Co-op Implements, co-op grocery stores including dry goods or pharmacy outlets, credit unions, and the Canadian Wheat Board. They voted CCF/NDP and belonged to the National Farmers Union. On the other side of the political spectrum there were those who supported a line elevator such as Federal, Pioneer, or Cargill, IGA groceries, McLeods hardware, and the banks. They voted for the Liberal and Conservative, "anybody but NDP coalition" and joined the Palliser Grain Growers.

By 1971 the Thatcher Liberal Government was already in trouble, as it had offended or neglected one labour group after another. People had finally accepted the public medical care plan as a tremendous success and finally blamed the insensitive political resistance of the Liberal-Conservative coalition. In response the NDP under leader Allen Blakeney offered a comprehensive and pragmatic election platform entitled "The New Deal For People." As a budding young university student at the time I was very interested in that progressive platform.

Albert, my boss, who grew up in a Conservative family was less convinced than I was but he would listen to my arguments. He agreed that the fight against public medical care by the Liberal-Conservative coalition was totally irresponsible and wrong. Albert also appreciated the numerous improvements in rural development in our province during the Douglas-CCF years including rural electrification, and rural telephones. As a result he was quite open-minded to new innovative government programs. To help matters for me the local candidate for the New Democrats, Hayden Owens a well-respected local farmer, implement dealer, and public school board member from Eston was a respected friend of Albert. Together Albert and I supported Hayden as did many others, so he was elected and the NDP formed government.

I was happy to work directly with Hayden on his campaign and fresh from a political victory I realized that the political bug had bitten me hard.

The difficult problem of keeping up with the huge capital requirements for operating a family farm has always been a significant barrier to entry into farming. Ever since the end of the free land homestead policies the acquisition of a land base for farming has been an incredible obstacle, especially for the new young potential farmers. Starting farmers purchase farmland if they are able to qualify for a "bank loan" and begin to pay interest and principal for twenty-five or thirty years. In the end they have paid three or more times the original cost of the land because of those bank interest charges. Then at retirement because their equity in the land is often their only pension plan, they turn around and sell the same land to the next generation, who begin to pay principal and interest for another twenty-five or thirty years. Again the credit agencies profit. Some say that is why the new combines have two luxury seats in the cab, one for the banker and one for the farmer. Since they are not making any more farmland, and it is obviously necessary for farming, the credit institutions continue to profit on the backs of beginning and established farmers, one generation after another.

The new government was ready to implement the Land Bank, a brave new program from its election platform. The function of the government backed Land Bank was to purchase land from parents at fair market value determined by the average of the most recent sale prices for land of similar assessed value. That sale would allow them to retire to the nearby town or city. Then the Land Bank would lease that land to the younger generation usually a son or daughter at a fair productive rent with the option to purchase and own the land whenever the lessee acquired the financial capacity. On the strength of those leases to age sixty-five with the right of renewal after that, many farmers where able to start farming, or purchase additional land to expand their farm and stay viable. It was a program that helped Sharon and I personally to build our own family farm.

In 1982 after ten years of increasing success, the Saskatchewan Land Bank was abruptly terminated by the right-wing coalition led then by the Progressive Conservative government of Grant Devine. He was able to portray the Land Bank as a dogmatic socialist intervention, designed to grab land from farmers at fire sale prices, with every intention to control the daily farming operations. This was clearly a regressive strategy to disregard the needs of establishing farmers for cheap political gain.

The "Devine Era" from 1982 until 1991 was a period of folksy rhetoric, philosophical inconsistencies, high spending, heavy debt, policy reversals and corruption. In addition that era saw much hardship for farmers due to drought and unstable prices for farm production. The Devine Government had eliminated the Potash and Oil Crown Corporations moving the government ownership out of those industries. It eliminated the Land Bank, the School Dental Plan, the Seniors Drug Plan, and sold off Highway Department road building equipment. All of that was consistent with a conservative non-government ownership political ideology.

However, building several new hospitals in sparsely populated rural areas, establishing a Gross Revenue Insurance Plan (GRIP) to guarantee minimum annual farm incomes, cooperating with the federal government on a Grain Price Stabilization Plan, enhancing the Saskatchewan Crop Insurance Program and accepting direct farm bail-out cash from the federal treasury, displayed a basic contradiction to conservative ideology. At the same time a high-spending, high-deficit, high-debt approach to government was digging the province into a deep-red financial crisis. None-the-less it was solidifying political support in the rural communities for the populous Progressive Conservatives who by then represented the coalition of the right. During that era the federal Progressive Conservative government of Brian Mulroney was in power from 1984 until 1991, which added federal resources to help prop up their high rolling provincial cousins.

The "Devine Government" was struck by internal legal and ethical corruption involving a few of its cabinet ministers. Their ongoing scandals along with the almost insurmountable provincial deficit and debt levels, did not sit well with voters across the province, who defeated them in 1991. They returned the NDP under new leader Roy Romanow. As strange as it sounds those negative factors associated with Devine's government did not result in any meaningful shift away from the Conservatives in their rural power base in Saskatchewan. That anti-right wing shift went to the cities and other urban areas resulting in a divisive rural-urban cleavage in our political landscape.

The new government was left in a near bankrupt financial position so in turn its actions became inconsistent with its own progressive rhetoric and ideology. Fiscal conservatism was now the NDP objective at least until the financial picture was improved. Hospitals were closed and services were cut or reduced, such as the elimination of provincially employed "agricultural representatives" who lived and worked in rural districts providing independent advice to farmers. They were valued for their agronomic and market knowledge, and were not employed by a farm-supply company seen to have a conflict when influencing farm business decisions. That one action was perceived as the government not understanding or caring about farmers. Also believe it or not, it was the NDP under Roy Romanow who passed the legislation necessary to allow the privatization of the Saskatchewan Wheat Pool. In fairness to the Romanow government that change in legislation was done at the request of the misguided SWP board of directors. However, the government could have stopped that historic miscalculation. A progressive political philosophy as consistent as the prevailing westerly prairie winds had finally shifted and blown away, in the stormy gales of financial crisis, pragmatism and hypocrisy. Nothing much was done during the Calvert years of NDP governance after Romanow to reverse that unfortunate drift away from the progressive ideals, but I believe, he did recognize and regret those previous policy diversions. By then the government had been turned over to the bureaucrats and party insiders. Progressive ideas could not flourish.

The Sask. Party Government defeated the NDP under the leadership of Brad Wall a devoted trainee with the Devine Era of reckless spending and political corruption. Wall has fallen four-square behind the pro-business corporate agenda championed by the Harper-Conservative government in Ottawa. Mr. Wall himself and his closest friends, I believe, see him as future prime minister and world leader material, ready to accept corporate admiration and lucrative pay-backs as he sells their right-wing agenda. That unified federal-provincial common spirit was clearly illustrated in the tactics those governments jointly employed in the elimination of the farmer controlled "Canadian Wheat Board Single Desk Selling Monopoly" in 2012. They have also established enabling policy for the privatization of the CWB by 2017. That initiative in my view is a total capitulation to the multinational grain trade. It is a retreat from the market security farmers with the help of progressive governments had fought hard for and won in 1935. A freedom lost. How could that happen?

For several years an "anti-CWB lobby" had argued that elimination of the monopoly would as sure as night follows day encourage the establishment of many value added wheat processing plants, which would spring up everywhere on the prairies. The theory behind that position seemed to be, that if farmers were able to compete with each other individually in the grain market—without a monopoly—the grain prices would be forced down low enough so a processing industry could profit. In that world, farmers would be subsidizing new value added business with farm income losses. By listening closely to the anti-CWB lobby, very early on, it was clear they wanted to have the farmer freedom for direct global sales at prices set through the existing powers of their CWB, turned over to the private grain traders. They wanted to go back over seventy-five years. Profits previously made collectively by a farmer owned and controlled CWB would be handed over to monopoly-powered, foreign, multinational-corporations. That was the objective and the eventual result. In fairness to those farmers who understood the reality and tried to save the functions of the CWB, they were up against a fierce, destructive force funded by a powerful business lobby at the doors of sympathetic conservative governments. The anti-CWB operational lobby was structured around incomplete knowledge, lies and diversion, supported by an endless supply of resources. Professional "spin doctors" without shame, carefully disseminated a regressive fraud.

I personally heard the federal Minister of Agriculture egged on by the provincial Minister of Agriculture boldly state that the Conservative Government after seventy-five years had come through on its promise to give farmers marketing freedom, and the control necessary to market their grain directly into the world market. What freedom and control was he talking about? The CWB had the marketing power for farmers to sell directly into the world market, and even set the prices on their product. That freedom would be removed, forcing farmers to sell indirectly to the world market through the multinational grain trade, which now sets the prices. The grain companies can now buy low from farmers, and sell high into the world market, transferring billions of dollars out of farmers' pockets into the international grain trade. Yet the Minister and the Grain Industry told us this is freedom. Farmers were fraudulently led to believe that eliminating the monopoly component of the CWB alone would not have any negative effect on returns to farmers. In essence farmers were told they could continue to get the benefits of a monopoly without a monopoly. It was a "spin" too good to be true. Members of the general public and some farmers, without a complete understanding of the economic facts, could not comprehend what the fuss was all about. They were sure they could have their cake and eat it too.

The Canadian Wheat Board Act required that a vote by farmers would need to take place, before a material change of this nature could be legitimately accomplished. However, the Conservative government argued that winning a majority in the federal election, fought on hundreds of other issues was democratic enough to justify these changes to the CWB. Unfortunately that election had used a strategy of political diversion to take farmers' attention off the CWB issue and put it on an emotional wedge issue of gun control. Prairie farmers hold a dominating affinity toward their guns, which they grew up with on the farm and any mention of government regulations for gun safety would immediately divert voter attention from grain marketing freedom to gun freedom. It was a brilliant political strategy used against farmers to kill their CWB. I say these hunter farmers shot themselves and every other farmer in the back.

The irony of this struggle is that the entire farm industry operates in a well-entrenched culture of monopoly decision-making by everybody else but the farmer. The multi-national corporations, which sell farmers their inputs, supplies, and services, all structure themselves as monopolies, duopolies, and oligopolies, to gain market power. Now without CWB monopoly freedom for price setting, farmers have lost the market power that every other business competitor in their market place has and values. The Saskatchewan Party government has taken action to protect Canpotex the international single desk monopoly that sells potash for fertilizer, a farm input cost, but it has worked with the federal Conservative Government to eliminate the Canadian Wheat Board international single desk monopoly, that sold wheat at premium prices for enhanced farmer income.

Too many farmers feel they are at one politically with giant agribusiness. They seem oblivious to the reality that giant monopoly corporations want to get as much as they can from the farmers' production. They generally don't understand that when wealth is taken from the farm sector and held by the railroads, grain handlers, traders, shippers and farm input suppliers, it is not available for any farmer no matter how hard that farmer tries to outdo his neighbour. Unfortunately, farmers too often see themselves at odds with each other, in a local competition to beat their neighbour, rather then in competition with the other parts of the food chain. Farmers generally have capitulated to the financial and philosophical pressures of the multinational corporations in a globalized economy, forcing them collectively to the very bottom of the multi-billion dollar food production and distribution system. There is currently a dangerous political climate of individualism which many farmers have adopted, often guided by good fortune, such as a lucrative inheritance and good health. They are easily encouraged by the free wheeling "survival of the fittest" business philosophy promoted in the corporate media. They see themselves more as multi-million dollar corporate owners and managers than farm workers and small business owners. Many farmers have not yet realized that in this highly competitive business world, they may have to trade-off some of their rugged individualism, and begin co-operating with each other, for a collective good, that in both financial and social value is better for everyone.

A problem I have noticed is that informed farmers are too full of self-doubt, while ignorant farmers are too self-confident. I believe to their eventual loss too many farmers have not informed themselves about the real issues and they have not had the courage to stand up to that well-orchestrated spin, designed and financed by the blind profit motive of corporate monopolies.

I sometimes say that farmers are prone to a devastating illness called "Busy Beaver Blindness." They are like ambitious beavers so preoccupied with whirling about building their home, that they become oblivious to the larger world. The micro-economic issues of farm yards, farm machinery, and farm production are getting more complex and time consuming, but the broader macro-economic concerns like market freedom, farm saved seed, global trade agreements etc. have also grown in importance. If the broader issues are ignored like transportation, marketing, insurance, investment and financing, the benefits of more precise production and increased yields can be eliminated. Producing more for someone else to profit in the food chain seems pointless. More production sold at lower farm prices may be better for the agribusiness industry, which profits on a volume basis but it pushes the farmer share of industry profits down.

Many farmers for reasons I don't ever understand have developed a combative non-union anti-worker psychology. That attitude confuses me because I think they have much in common with unionized "Blue Collar Workers" who like themselves are welders, plumbers, carpenters, mechanics, machine operators and physical labourers. In my opinion farmers are very similar to middle class workers even if they do own a large farm corporation, because they are not in the league financially or socially with the ownership and management of large global companies. None-the-less too many farmers do not support policies or collective action helpful to middle class workers. They think of a well-paid worker, especially a unionized worker with benefits a farmer may not have, as their enemy. Traditionally farmers have not offered their hired help much in the way of compensation, benefits or job security. Unfortunately they see themselves in the context of owner enterprisers, and see farm workers as a commodity in a cost cutting culture, to be exploited for profit. That attitude may need to change as workers become more in short supply.

I feel there is a dangerous degree of complacency and apathy, generated by a false sense of security that historical progress made in terms of economic and social gains can never be lost. Folks in the farm communities are in a state of confusion and despair as they confront the rapid acceleration of off-farm migration and the concentration of farms into the hands of a few family elites.

Future Family Farms (FFF)

Looking forward I sometimes think about how the (FFF) will be organized and be managed and I wonder how our future farmers will relate to each other, their community, the labour movement, the co-operative movement, governments, crown corporations, multinational companies, the global economy and the environment. My experience tells me that complex social, economic, democratic, and environmental matters are as relevant in our farm communities as they are anywhere else. Issues such as climate change and the pollution of our environment should actually mean more to farmers than anyone else. Who could be more connected to the land and the ecology? One should not make the mistake of assuming in this information age that our rural communities are less informed on these issues. In my opinion farmers are generally capable of designing and implementing their own future. The question remains will that be a future that improves our society?

Some of our family farms have already grown to tens of sections (640 acres per section) of land, and they continue to expand. Some farms crop over ten thousand acres. Our farms are being concentrated into the hands of very few privileged families with the accumulated wealth needed to continuously expand their farm. Ownership and investment in those large farms has expanded to family members not living or working on the farm and non-family investors who transfer wealth from other occupations and locations back to the farm. Therefore the farm market, especially the land segment, is not a closed economic system where prices to buy land directly reflect the productive returns from that land. Land and machinery prices have often greatly exceeded the market realities.

Rising costs have reduced the profit margin per acre, which means more acres must be acquired to make a living. In the absence of a Land Bank transfer agency, many young farmers do not have the capital to get into the business so the industry is turned over to fewer and fewer wealthy farm families. Farms that have moved away from raising livestock have freed up time in their work schedules to grow crops on larger scale operations. They have also developed into sophisticated corporate enterprises. For the most part the decline in farm numbers has been felt more by the mid-sized traditional grain farms. The main contributing reasons for this have been the new developments in mega-technology as well as new corporate farm organizational and management structures.

These new "Future Family Farms" are not the "Traditional Family Farms" which were often run as mixed farm operations struggling to survive with a relatively stable population attempting to hold the rural communities together. Now with the arrival of large "Elite Family Farms" the taxpayer citizens not on the farm have become less interested in assisting farmers with any government support programs. Especially when they see large farm corporations acquiring vast amounts of property and wealth and sending many of their farm neighbours off to compete with them for jobs in the city.

Perhaps I digress but whenever the subject of farm size comes up it reminds me of the time in a restaurant when I overheard four farmers talking about the size of their farms. One guy was from Prince Edward Island, another guy from Alberta, a fellow from Texas and someone from Saskatchewan. The farmer from PEI was saying his farm was so big that if he started walking at sunrise, by nightfall he would just get to the other side. The Alberta guy said if he started riding his horse at sunrise he would just get to the far side of his farm by sundown. Then the Texan bragged that if he got in his pickup truck at sunrise he wouldn't get all the way across his farm by night. That was when the Saskatchewan farmer said to the Texan "I know just how you feel because I once had a pickup truck just like that."

I suggest that two fundamental economic principles have collided on the Canadian prairies. One is the concept of "economies of size" and the other is the opposing counter balance of "diseconomies of size." As profit margins are reduced for each unit of farmland held, the economies of size principle says you should add more units of production, in order to maintain or increase your overall profit from your labour and management. The underlying theory is that as your farm gets larger you can stretch fixed costs, many of which would also be necessary for smaller operations, over a larger enterprise reducing the per unit costs, hence generating more profit. However the law of diminishing returns or diseconomies of scale predicts that as you increase the size of your farm and reap the added profit, you will come to a point of diminishing returns. At that point adding another unit of size, adds new costs more than the increased amount of profits from adding that new unit. The reason being that at some point you will not have the capacity in terms of capital, labour, expertise or time, necessary to operate the new larger farm. More investment in those components will be needed to survive. So where does that leave the less fortunate farmers and their future family farms, if they don't have the resources to expand, you might ask?

Older folks have responded to these new farming challenges by accepting them as a signal to retire to a town or city closer to health facilities. There is also the constant desire by some family members for advanced post-secondary education, which has lured young people away from the farm communities. They go away with their farmer optimism in place, with a constructive view toward taking on a new career. Some folks are just not interested in the farming business or do not have an aptitude for farm duties. As we say in the farm lingo "some people are wired for farming and others are not." There is also the reality that for some the grass is just greener on the other side of the fence. Farm life, no matter how lucrative or rewarding it is, or could be, remains restrictive in terms of mobility, since you are required to be closely attached to your farm at least for certain periods of the year, if not all the time. Many farm raised young people are attracted away by a simple desire for mobility. Health issues can also obstruct entry into the farm business. Sometimes a young person raised on a farm meets a life partner who is not interested in the virtues of farm life, so they move away. These realities of life can be a sad awakening for a senior farm couple, who feel they can provide a farm opportunity that a son or daughter cannot refuse, but they do.

Ironically there are some notable anomalies to these futuristic trends. Contrary to the dominant expansion mode some smaller farms will continue to operate. As a matter of fact, there has been an increase in very small farms as intensified operations growing specialty crops such as fruits and vegetables. Smaller organic farming operations without the use of fertilizers and chemicals have also been on the increase. They have responded to an expanding niche market and have actually reduced the unit costs for farm operations. Cattle and other livestock production in a traditional mixed farm operation has been the most common form of diversification over time. Many of them will continue. Running a seed cleaning plant or some other value added venture has also been combined with the farm operation. That will carry on. Working on the side at occupations such as carpenter, plumber, electrician, welder, painter, or general labourer either as part of their farm operation or in town for someone else, will continue as usual. In addition many farm people hold down full-time jobs in a nearby town or city while running their farm as a sideline. That opportunity for diversification will continue but will obviously depend on the prosperity of the farm communities in general.

Then there are those who need to move and live outside the local farm area, sometimes, long distances away from their farm, in order to earn income to keep the farm viable. Sometimes those arrangements are just for the winter months and in the case of a straight grain farm, the farmer can return in the spring something like a working snowbird. Unfortunately some farms will disappear to become part of a larger farm, and years of work and cost will be gone forever. However for some their farmyard will survive as someone's acreage complete with home, shop, shed, barns, fences and all the services. One way or the other many farms will adjust as best they can and continue to thrive.

My good friends "Joe the Boss and the Preacher" and "Jerry the Finance Manager" from a Hutterite Colony west of Saskatoon, meet with me and a few others, every so often for coffee, when they come to Saskatoon on business. Usually Fridays. Sharon and I have been out to the colony on their invitation to socialize and learn how they manage the farm, church and school. There is much to be learned from how the colony provides for itself with highly intensified business operations. I believe they will survive and grow.

Labour is in good supply as everyone has a job and is supported to become extremely proficient and educated at the job they are selected to do. From an early age, after school and on weekends, the children learn to assist with all kinds of farm chores. They have a sophisticated, giant, state of the art machine shed for mechanical and servicing work, which includes a general shop area, a modern body shop, a full service carpentry shop, as well as kitchen and business office areas. Somebody is the farm boss, somebody is the garden boss, somebody is the cattle boss, somebody is the hog boss, the dairy boss, the woodworking boss, the plumbing boss etc. These folks have the best of modern farm machinery and service tools.

The women and girls are in charge of the cooking, sewing and household chores, as well as the garden. It is not my view that the women consider themselves mistreated or without independence in life. I remember asking Ester—Jerry's wife—what she thought of Jerry's experiment to plant some radishes in the fall to get the jump on spring time. She said: "That was the stupidest idea he has ever had." There is no doubt that these ladies like to come to town, any town, and shop. They are a solid part of a happy life, with the ordinary ups and downs of any other family.

I guess maybe because of the collective ownership of the colony and the top-down almost dictatorial decision-making structure, right down to the similar black and white apparel, from shoes to the hat they wear, we are tempted to not see the common humanity in the loving family life of the Hutterites. My friend Joe will tell me as he points to his clothes "We sometimes have drunks, thieves, liars, scoundrels and assholes that hide behind these black clothes—but not many, just like everywhere else in the world." Then he will explain to me how tough it is to discipline members of the colony when it is necessary because they are often in his own family. He affectionately calls me "Danny Boy" when he speaks directly into my eyes knowing that his words will penetrate right down into my heart. Joe says: "Danny Boy I know you have worked hard in order to build up your farm and I know you own much of it, because you have it rented out. For me the minute a baby is born on our colony that little baby owns as much land or property as I do. I don't own anything but it is my job to run the colony for everyone."

Both Jerry and Joe are devoted family members proud of their wives and kids. One day Joe candidly told me about the sadness and turmoil that hurt his life when he lost his first wife who was stricken with cancer, soon after they had a young son. He told me how loving his wife's younger sister was to devote almost all of her time to care for her fading sister and Joe's son. Eventually that younger sister of his first wife would become his current wife. Jerry has also had a battle for his health and knows his way around the public health care system, which he admires. Joe, Jerry and I share an affinity for schools and the educational success of all our children. However the colony only provides schooling up to grade eight in cooperation with the public school division. Joe and I disagree as I often tell him that I believe his colony should increase the schooling for its' kids up to grade twelve graduation. I am still working on that objective knowing the colony has bright young learners and should have that option in life. Joe has been brutally honest with me as he says there is a fear that additional education may eventually attract the kids away from the farm colony. That he says could destroy, their traditional simple but secure way of life. He doesn't completely rule out such a change if it is done in the best way. Under the present lifestyle food, shelter, furniture, clothes, employment and retirement are provided. Jerry the finance boss even makes sure a reasonable amount of booze is available to each family unit provided it is not taken to excess and interferes with work, health or family life. I believe that the culture and direction taken by each colony depends on the chosen leadership.

Any discussion about the Hutterites needs to emphasize how much they are willing to support their neighbours, and the surrounding communities. Not just by selling them garden produce at the farmers' market, or chickens and other meat directly to customers, but by helping out when someone has bad luck and needs a hand or a community project needs food, manpower and machines. If there is ever a fire, the people from the area know the colony is always ready to help. In addition to the normal farm operations, which are done very efficiently the colony takes on various outside contract jobs such as custom auto-body repair or putting large machinery together for farm equipment dealers. Direct selling of garden products at the local farmers markets is also a lucrative business enterprise they manage. There is no doubt in my mind that because of the business savy and the intensified nature of these Hutterite Colony farm operations, they will adjust and flourish into the future. We need to accept that they really are viable family farms.

The historic contribution of all farm women in the fields, in the barns, in the gardens, in the homes, in the farm offices, in the communities, in politics and even at off-farm employment has to be recognized as a key reason for any family farm success. Our own farm is a good example. I recently said to Joe and Jerry at one of our coffee meetings: "Nearly thirty-five years ago I said to my potential wife Sharon, that if she married me she would never have to work another day in her life—so she will be retiring at the end of this year" then I boldly stated "some things are worth lying about." Joe put on his preacher's hat, pointed to the sky and said "No-no Danny Boy."

The reality is that many traditional family farms as romantic as they are perceived to be, are quickly going by the wayside. Unfortunately several rural communities are going with them. Too many middle-aged farmers with irregular and often low farm incomes, feeling enormous regret and much pent-up emotion, have been or will be forced off the farm by these changing conditions. This whole process of forced exodus is not a harmless evolution and it has or will have its personal, social and political consequences. I call them "uncosted disbenefits."

New, younger farmers are difficult to find. Those who have stayed or have come back into the farm business more recently are generally from a wealthier established family, with the benefit of inheritance, or other financial support. This changing of the guard has also manifested itself in a changing political environment. This new neo-conservative population, without the personal experience of dealing with devastating perils imposed by the land, the weather, world markets, banks and other multi-national corporations, seem to take their farming status for granted. It is their entitlement. These folks often see themselves as bullet proof and think nothing can happen to them that their new found wealth, or their parents resources will not take care of. Some even think a period of general misfortune may even be good for them personally as they displace yet another neighbour and divide the spoils. The social and economic institutions designed, implemented and supported by the "progressive movement" in the past, I suggest are at risk going forward, because these new farmers have not yet experienced the brutal forces, which motivated their development. The romanticism of rural farm togetherness, once demonstrated at a barn raising or branding, has all but gone, replaced by a new rugged individualistic entrepreneurial climate. There has been a dramatic shift from a "we to "me" attitude.

Ironically farms that have very successfully devoured each other, and in many cases grown to the giant sizes they are; now face a serious shortage of farm workers. That scarcity has become a severe problem. The displaced farm families who have left the farm community are now needed back on the farms. The chickens are coming home to roost for these new landholders as their labour requirements have increased significantly and the cost of labour is now a serious diseconomy of scale. Serious adjustments in employee-employer relations on the farm will need to take place. Job security along with improved compensation and benefits will be required to attract back the necessary workers. In my view temporary foreign worker arrangements will not be secure enough to supply the stability and quality of labour needed. We need to repopulate the rural areas, at least in the larger centres close enough to the farms to commute for work.

The Future Family Farm I believe will continue to need the services of welders, plumbers, carpenters, mechanics, cement workers, roofers, butchers, bakers and candlestick makers. The difference is that the traditional farm family once would provide all of those services from within its own extended family. On the new farms almost all of those services will be provided by external contractors or businesses located in larger centres often as far away as the larger cities. Farmers who once used the off-season for machinery repair and maintenance will take that tractor or combine to town to have it "green lighted" as they say, ready for the next season. Farmers will hopefully find these needed services in a local town that depends on the delivery of those services as a reason to survive. A town based on that service structure will then be able to provide the schools, doctor services, legal and accounting services, and some of the finer things like fast food outlets.

I suggest there is a lesson to be learned for the (FFF). The recruitment and retention processes in response to the new farm labour requirements must provide a more inclusive and respectful approach to farm workers for these future farms to survive. New collective labour-management agreements will need to be negotiated to offer farm labour stability and the economic sustainability for our rural communities. Working on a modern Saskatchewan farm should provide as much job security and income stability as employment at the local co-op, a grain company, the school division or local government. Organized farm workers, unionized to secure and maintain wages, pensions, and safety conditions, will be part of the new environment. Establishing support services for a new well-paid working class will provide important opportunities for the local business community.

The financial and human resources required for future farm operations such as land, machinery, technology, management and labour I predict will no longer be supplied entirely by one individual or even one family. Farmers, who previously farmed their own farm will still have ownership of their land resource but will rent it to the larger farm operation. The agreement may be a cash payment rental arrangement, or a crop share rental agreement. Sometimes the landowner will just get the larger farm operator to farm his land on a fee for service arrangement where the landowner manages his own farm and pays for whatever work is needed, and gets the crop himself. The owner retains his farming status as manager. There will be a growing concern that the large farm units will not be able to operate the combined farms with the quality of care that the previous farmer had exercised. Therefore farm families entering or continuing in these amalgamated operations may become more selective and willing to renegotiate or change the arrangements from time to time. Farmers may begin to move their assets from one farm arrangement to the next to demand quality and efficiency. They may even seek out a relationship where they will rent their farm to a larger unit farm if they can work on that larger farm or even manage all or part of the future farm.

Some of these modern financial arrangements will involve rented machinery. Equipment may be rented from a private individual, a corporation or a co-op formed to invest in machinery for rental. Farmers will have an opportunity to get together as a partnership to solve a variety of these service issues. For example, farmers can get together and purchase a transport truck, and hire a capable driver who may be a previous farmer to operate and service their in-house trucking business.

Seniors retired from other occupations could enjoy the farm life for a while by helping out on the farm and making some additional income. Students attending post secondary education could find a summer job on the farm. Relatives or friends of the owners and managers may invest outright in land with the knowledge that they can have it farmed by a farming consortium and over time collect a return on their investment. Other individuals, corporations, cooperatives or even government agencies such as pension plans, will be able to invest in land or other farm assets with a mutual agreement to lease it to the future family farm.

I think in some cases a new more sophisticated and formal business arrangement based on the mutual benefits of cooperation will need to be designed and implemented. I believe that Federated Co-operative Ltd. based on the strength of its network of modern retail co-ops should have an important role to play in the regeneration of a progressive farm economy. Farmers and governments should look to FCL as a potential partner in an attempt to regain their presence in the grain handling, storage, transportation, and marketing functions of agribusiness. FCL will need to exert its leadership potential, reinventing itself as a co-operative not just continue to emulate other established corporate business models. It will need to expand its horizons beyond the oil industry.

Governments need to step up to the plate and do their part for the farm economy. They need to produce solid regulations and controls on plant variety registration and genetically modified seeds (GMOs). Farmers need protection for farm saved seed and public plant breeding to maintain their freedom that may be lost to monopoly patents held by non-farmer corporations. Governments need to implement effective quality control standards for farm inputs such as fertilizers and chemicals.

Strong performance regulations for the orderly marketing of farm production through the purchasing, handling, storage and delivery systems are required. Quality control throughout the system from farm to end-user is essential for the stability of our markets. In order to accomplish these objectives, governments will need to provide the necessary resources for the inspections and scientific analysis required. All of this cannot be accomplished if governments rely on the industry participants to police themselves and if governments do not provide the resources to support independent evaluations and controls, which are essential.

I believe the majority of farmers will soon demand that the federal government revisit the decision to eliminate the Canadian Wheat Board single desk selling monopoly. Billions of dollars in lost premiums and the untenable disruption to the orderly marketing system, will force all farmers and the general population to reconsider that misinformed decision. Importing countries previously relied on our higher quality grain to mix with lower quality grains to bring them up to minimum standards. Now that those CWB quality assurance mechanisms from the orderly marketing system have been removed, I say it is only a matter of time until importers apply irresistible pressure on governments to force the grain trade to improve quality control standards, and improve the regulatory and inspection processes. The best way to do that is through CWB orderly marketing. Recently six of seven farmers elected to the new Saskatchewan Wheat Development Commission, clearly took the position that removal of the CWB monopoly was a mistake.

Farmers will need to relearn that farm cooperatives working with progressive governments have given farmers effective market power in the past with excellent returns. Farmers in the area of wheat marketing actually had so much power that the multinational corporations who are our competitors for farm profit share used their dominant financial and political power over several years to destroy our advantage. Now they have succeeded and farmers will pay the price, until we can come together in farmer owned and controlled co-operatives to regain that market power. With or without government support, that will need to be done. History tells us that the original CWB was dissolved from 1931 until 1935, for financial reasons, before it was restored. I believe it can be done again if farmers, our communities and our elected politicians demand that it be returned.

Canada needs to establish an international trade policy that strives to protect and improve the interests of workers and farmers not just CEO's and shareholders. Shareholders in the equity markets are only one component of the stakeholders reliant on the farmer and the food chain. Any trade agreement that eliminates the power of our sovereign nation, to democratically implement policy in the best interests of our farmers and our country, should be revoked.

Farmers will learn the hard way that the true fight for farm survival is not with each other and it never has been. They will come to understand that as individual farms they will not be able to buy their way to the head of the line in spite of what they may have been promised during the CWB debate. That false sense of individual security will vanish as agribusiness strips farmers of their share of the growing food chain wealth. Farmers will learn that what is good for the general economy and agribusiness is not always good for farmers. Inexpensive food for all Canadians, and large industry profits, need not come at the farmers' expense. Unfortunately, at only three percent of the population farmers do not a large political movement make. However going into the future our farm economy, a safe and stable food supply, a sustainable environment and the democratic control of our country are issues that will affect us all. Granted some folks are affected more than others but as Benjamin Franklin once stated: "Justice will not be served until those who are unaffected are as outraged as those that are."

Farm Life

by Dan Danielson

Farm Life is a happy child of nature,
With golden fields and friendly towns.
Offering peace so calm and secure,
Nature smiles more than she frowns.

While working in our new farmyard,
I stop to recall my humble past.
And remember progress comes hard,
Farm success doesn't happen fast.

Our farm feels the forces of inflation,
With rising costs and high interest rates.
Unsure market prices cause frustration,
While cruel bankers set collection dates.

The prairie scene of Saskatchewan,
Will have lost so much of its charm.
And rural happiness will be gone,
If someday only the rich can farm.

six

EDUCATION AND EMPLOYMENT TO SUBSIDIZE THE FARM

As a young guy starting high school in Eston, I recall seriously contemplating my future, probably years before most young people burden themselves with such complicated issues. For one reason or another I was more than a little bit disturbed as to why some families had more than others. The concept of inequality has always bothered me and no doubt has helped shape my attitude toward life. It occurred to me early on that education is a great equalizer in our society and that moving up the unmentionable "class structure" is possible with success in education. I was convinced by examples of successful folks I knew, and by my parents, that a process of life-long-learning which blended formal education with work experience would pay off down the way. I was definitely helped to arrive at that conclusion by my mom who was a dedicated teacher and my dad who dwelled on the need for hard work. He probably promoted education as much as anyone because he had not gone past grade-eight and he wanted me to do better. He was educated in his own way when he saw much of the world as a soldier in WWII. That was a somewhat unique education that I always admired.

From my graduation in 1967 onward, I knew that my career needed to involve off-farm employment because my family didn't have a large enough farm to hand over to me and still have the wealth to retire. There are three siblings and myself in our family and we all understood that even after the eventual passing of our parents, only four-quarters of land would be shared equally among us. As a result by necessity my eventual career evolved into a double-barrel undertaking, made up of various employment opportunities assisted by my education on one side and growing a family farm on the other side. Both would be a formidable challenge worth accepting.

Over the years I was able to achieve the following formal education and acquire related work experience skills.

Education:

1. Master of Education Degree (M.Ed.) in Education Administration (1988)
2. Post Graduate Diploma (P.G.D.) in Education Administration (1986)
3. Bachelor of Arts Degree (B.A.) in Economics (1984)
4. Bachelor of Education Degree (B. Ed.) –Teaching Certification (1972)
5. Advanced Certificate in HR Management (2000)
6. Coursework:
    a. Canadian Securities Course (2011)
    b. Canadian Securities Conduct and Practices (2012)
    c. Life License Qualification Program (2013)
    d. Saskatchewan Life Insurance Council Bylaws Exam (2013)
    e. Financial Planning (2013)
    f. Professional Director Certification (2015)

My main areas of interest and expertise are education, economics, agriculture, human resource management and financial management. I have a passion for the theory and practice of effective organizational development, especially improving the relationship between policy and administration functions. My academic thesis for my Masters Degree was researched and written on the concept of policy making. I feel comfortable working with rural, urban, and northern people, and I am very proud of my success building effective partnerships.

Employment

My employment experiences have been many but usually of a short-term nature fitted into my farming career. While I wanted to experience the city life and various off-farm employment opportunities, I never really strayed very far from my home area or found many jobs totally unrelated to farming.

It took a while for me to realize that what I really wanted was the best of both worlds. Fortunately for us in Saskatchewan the largest cities of Saskatoon and Regina are situated in farm country. Therefore with some confusion, some adjustment, and lots of hard work, I tried to experience both. I occasionally hear folks from the farm say they could never live in the city, and I regularly hear city folks say they could never live on the farm. I know that given the opportunity I prefer to live in them both. Usually the farm in the summer and the city in the winter.

The Collins home where I lived

Albert and Dorothy Collins

Money was very tight in my day for young people and most everybody else. For me with a farm background and being mildly competent at all sorts of basic farm duties, and a burning desire to spend my summers home at Tyner; I was delighted to be offered a job on a neighbour's farm. It was a job I could begin early each spring and work as long into the late fall as I was available. At $2 per hour it was not the highest paying job I might have found but any other option probably would have taken me out of my home community. As a committed ball player I liked the opportunity to stay in the area and play fastball. I liked farm work and was treated very well by Albert Collins, my boss and his farm family. Dorothy, his wife was a lovely lady who treated me like one of the family. Their place had a large majestic "Eaton's Catalogue Farm Home" where I lived during the summers, complete with a well-stocked food pantry. An additional perk was my valuable and much appreciated access to the farm fuel tank for my thirsty car. I had the boss's permission for the gas as long as I would risk using farm purple in my 1965 Chevrolet Impala, two-door hard-top, which was white on the exterior and had a hot red cloth interior. My car thrived on the purple stuff and my wallet also felt the pleasure of free gas, but I was reasonable with my consumption. You might also like to know that my sporty car had the talent to go down the railway track, without even letting air out of the tires, if you didn't move the steering wheel.

My parents who were quite interested in my future success were visibly happy that I fit in so well with the Collins family. It was a family with whom they continued to be devoted friends. Both the Collins Family and the Danielson Family by then had taken up part-time residency in Saskatoon for the same reasons, to be closer to post secondary education for us kids, and help subsidize the farm. Based on that split residency practice the label "Suitcase Farmer" was born. In the beginning I believe the term was intended to be a pejorative expression, adopted by those who felt my dad and Albert had abandoned the farm for the winter. For me, it became a very positive expression, meaning the freedom to expand ones horizons beyond the farm scene but holding onto farm life.

Albert found his interest in city life growing and wanted more time there and consequently less time on the farm, although he was a good farmer and wanted the farm to continue to be successful. Because I was available and interested in as much time on the farm as possible, our working relationship continued to thrive for several years. Some weeks he wouldn't come to the farm until Monday afternoon or even Tuesday morning and then be ready to head back by Friday noon or earlier. He was helping his family establish a "Hobby Shop" business in Saskatoon. I was proud of the fact that he relied on me to handle the farm duties in his absence and he would tell others he was fortunate to have me as his foreman. It was a boost to my confidence and motivated me to take on more responsibility and do a good job.

Whenever one of his own four boys was around or when my younger brother was available, Albert would also hire them to help us out. They had other career preparation interests on the go but came to the farm when they could to earn a few extra dollars. In order to keep us all working and make money to pay us, Albert would arrange to custom work as much land as he could find. At one point, way before the days of large corporate farms, we were farming about five sections (about 4,000 acres). We did it with two eighty horsepower tractors, one twenty-foot discer, sixty-feet of harrows that could be converted into a sprayer, two small Massey 92 combines and two grain trucks. Those were the days when you ploughed half of the farm acres for summer fallow and cropped the other half. We had two sixteen foot deep tillage cultivators, and two twenty-four-foot rod-weeders, that we used to keep the summer fallow clean. That venture meant long hours, sometimes though the night, but we got it done and more hours meant more pay. Perhaps I didn't appreciate it enough at the time but I now know Albert was doing this for us young people, wanting to help us work for an education and a better future.

My boss taught me a useful lesson on business agreements. Albert would say that written agreements or contracts are important, however, he suggested that agreements only succeed if they represent relevant and understood benefits for all parties. He taught me that it is the recognized mutual benefit of any agreement that holds it together not the paper document. If there isn't a commitment generated by the mutual satisfaction gained from the business relationship, then one or the other of the participants will want out and may not live up to their side of the bargain. That insight was very true for the unwritten agreement I had with him. Our agreement gave me the opportunity to earn income in order to go to university, yet stay in my home area in the summer to play ball. The agreement also satisfied his need to hire experienced farm labour to help his farm continue to succeed. Our arrangement was a fine example of the kind of mutual agreement he spoke about.

In 1971 the newly elected NDP government set out to implement the "Land Bank." Hayden Owens, our new MLA, was committed to the success of the Land Bank and he knew the new government wanted a few university graduates, with farm backgrounds and a belief in the Land Bank concept, to help set up the new program. Hayden was convinced that I was needed on that implementation team and he knew that I was about to graduate from university in 1972. I was told later that Hayden made many trips to visit government ministers to make sure I was one of the first twelve people hired to get his pet-project going. My job as an interviewing counsellor was to sit down with farm families, to explain how the program could work for them and help them through the process to make an intergenerational farm transfer. It was one of the most rewarding career experiences anyone could have. I was able to observe and experience up close the degree of satisfaction the Land Bank brought to many farm families.

As I did my job, I would find time to keep Hayden informed of the great success his new land transfer baby was having. Together we would chat over a coffee in the legislature coffee shop, enjoy our common friendship, and express our mutual satisfaction with the Saskatchewan Land Bank. That program assisted over twenty-six hundred family farms over a ten-year period.

My employment at the Land Bank went very well but it needed to end when I decided to rent land myself from the Land Bank. The neighbour who I was already renting land from had a sudden heart event and decided he wanted to sell his farm. He offered to sell to me and I wanted to buy it, but at age twenty-four with no equity for security, I couldn't arrange the financing for an outright purchase. However, with the assistance of the Land Bank, a program I knew worked very well, I could continue to rent without losing my start at farming. My landlord was able to also get what he wanted from the sale of his land.

Following my resignation from employment with the Land Bank I was asked by the provincial Department of Natural Resources and Tourism, to lead a team to conduct a study documenting the amount of Saskatchewan farmland and recreational land owned by foreigners and corporations. It was my first opportunity to be a manager and to make a mistake that got me into hot water with senior management. I had decided that the best way for my team to obtain the ownership information we needed, was to get it from the rural municipalities off their tax rolls. As a result I designed a rather extensive but complete document which we asked the administrators of all the rural municipalities to fill out for us. Members of my team would follow up with those administrators if necessary with a visit to help them out with the process. I had unilaterally made that decision without senior government approval and I agreed to pay each administrator twenty-five dollars for completing and returning the form. My reasoning was that if they completed it fully we would not have the cost to travel out to see them. As a new manager I didn't know what kind of trouble that decision would stir-up for myself.

One day I was unceremoniously called into the Deputy Ministers Office to meet a couple of Assistant Deputy Ministers, a government Treasury Board Official, and Dean Jake Brown from the College of Agriculture, who was familiar with the kind of research project I was managing. Much of the meeting consisted of government folks lecturing me about how things had always been done around there. That scolding process I assumed was intended to shake the daylights out of my self-esteem. And it did.

Finally it was time for Dean Jake Brown from the University of Saskatchewan to speak. Mr. Brown said he had not seen a study of this type done in this way before, so for that he said I was guilty of non-compliance with tradition. However, he also said it is hard to argue with success. He knew of the very high percentage of responses we had achieved, and stated that our rate of success was the highest he had ever seen. Besides, he emphatically noted the time and financial savings accomplished by my new way of doing things. On that basis, he recommended that the government financial people should approve my payments, and he advised that the method my team and I had designed and implemented should be used for all similar projects in the future. That is exactly what they did and my reputation recovered from mud hole to mountain. So did my confidence in my work and much to my surprise I learned that I was not going to be fired.

I guess right then and there, I confirmed in my mind that just because one method or approach to meet your objectives has always been used in the past, it does not mean that there is not a better way. I suppose I learned the hard way, but ever since, I have been rather rebellious when it comes to breaking from tradition when it appears that a new way is better. That experience taught me a new form of "progressive aggression."

A day or so after my run-in with my superiors in government I stopped by Ross Moxley's office at the Land Bank to catch up on the office news. Ross was Vice-Chair of the managing commission and one of the people I had worked for during my time at the Land Bank. Little did I know how much my work-life would change as a result of our brief discussion. We exchanged greetings and a comment or two about recent life events, then Ross said before I barely sat down: "how would you like to be the first Director of the new Saskatchewan Farm Ownership Board?" Jokingly I replied: "That would mean more pay I guess." I had known that the main purpose of our Foreign and Corporate Land Ownership study was to form the basis for a law and related administration to restrict the amount of land that could be owned by corporations and foreigners. It was intended to reduce the foreigner and corporation competition for farmland making it more affordable for new farmers as well as smaller family farms that needed to expand.

Ross went on to explain his own career situation and how it had prompted him to ask me that question. He told me he had actually been offered that directors opportunity with the Farm Ownership Board and in fact he had accepted. However, in the mean-time something else had come up for him. The government had just begun to establish a new Department of the North based in La Ronge. The current Deputy Minister of Agriculture was moving up there to become the first Deputy Minister of that new initiative and he wanted Ross to be his Assistant Deputy Minister, working in intergovernmental affairs, along with the Federal Government, on northern issues. Ross told me he had always wanted to work and live in the north and this would be a promotion for him as well.

Then he told me that if I was interested in the Farm Ownership Board position, he was sure he could convince the Farm Ownership Board members, most of which were familiar with my work and education, including Dean Brown, to offer me the director position. To make a long story short, I would accept an acting role for the winter months and we would discuss an extension in the spring. That is how I officially became the first Director of the new Saskatchewan Farm Ownership Board.

My first task with the FOB was to arrange to relocate the administrative offices out of Regina from agriculture's planning and research division in the Walter Scott Building, up to Saskatoon into the tenth floor at the top of the Bank of Commerce building. In addition I was in charge of hiring the administrative assistant staff and a research officer. Harley Olsen who I had met at planning and research came on as our "Research Officer."

Within one year as much as I loved my work, I would be faced with a huge career decision. Again torn between the best high paying job I could possibly have in the city at that time, and my return to the farm, I had a life changing decision to make. They did offer me the job on a permanent full-time basis in the spring but I declined in order to work to improve our farm. Probably not the wisest financial decision I would ever make. None-the-less I decided that I would continue to seek out meaningful winter employment but return to my farm each summer. Again this was my idea of having the best of both worlds. Therefore I would be involved with hiring a new permanent director to replace myself.

In 1975, the issue of rail line abandonment was front and centre in our province as both the CPR and CNR were attempting to get government approval to close down many railroads. In response, the Federal Government engaged a prominent Saskatchewan Judge, Justice Emmet Hall, to lead a Royal Commission on Grain Handling and Transportation across the prairies to hold hearings, giving rural people the opportunity to make representation for the retention of their railroads. That process became known as "The Hall Commission Hearings." Often the hearings would be held in a relatively small town hall with participants and supporters coming out to present their views in "briefs" to members of the Commission. Sometimes the communities would go all out to host the event with local entertainment and a lovely home cooked meal, complete with my favourite part of the whole day, eating home made pies.

In order to assist the communities to prepare for the "Hall Event" and help them write their presentations, the Saskatchewan Attorney General's Department hired twelve advisors to pan out across the province. Roy Romanow, as Attorney General in the Blakeney government, decided to hire six recently retired top executives from the Saskatchewan Wheat Pool and six young university educated farm kids to team up as provincial advisors. I was fortunate enough to be selected at twenty-six years of age as one of the young people to be paired with an older retired grain industry executive. I was paired with a very personable and experienced fellow by the name of Able Toupin, who went back all the way to the early days of the Saskatchewan Wheat Pool. He had served in the prestigious and influential position of "Fieldman" later to be called "District Representative." Those folks in the early days were instrumental in signing up farmers to support the Sask. Wheat Pool, the Canadian Wheat Board and the Canada Grain Commission. It would be hard for me in this short writing to fully cover how much I learned from my senior partner. He was a man with an amiable personality, a good sense of humour, impeccable people skills, great farm experience, abundant knowledge, and an unwavering commitment to the co-operative principles.

It was on this interesting job that I saw first hand why somebody came up with that famous little proverb: "if you really need something done go ask the busiest people to take on the work." Those were the dedicated folks who would make time to do the job. We would make three or four trips into each of the affected rural communities prior to their assigned day, when they would meet and present to the Hall Commission. The first trip or two would involve talking to people from the Town Council, RM Council, Community Association, National Farmers Union, Sask. Wheat Pool Committees, Church Council Members, 4H Leaders and Members, as well as any other community activists we could find. On our first planning day we would present the group of community representatives with a guideline document, which I had prepared. On the next two or more visits, we would go over what they had put together and fill in any of the gaps, as well as organize and rehearse for the "Big Show." Then as advisors we would make sure we were there to help out and provide moral support on the day of the formal Hall Commission Hearing. It was an important occasion because the results of that process would impact on the communities for a long time.

As a young farmer who had already worked at a number of farm-related government jobs, I was impressed with how well informed, organized, and articulate the representatives of The National Farmers Union always were, whenever I came across them at meetings. They especially impressed me with their knowledge and professionalism when they helped develop the community presentations to the Hall Hearings. Therefore as a farmer I decided to become a member of that organization. In some ways the term union in the traditional sense did not really apply to the NFU, because it didn't really have a massive membership, with a realistic option to strike or withdraw services in any significant action. However it did have a very meaningful and influential function as a research and educational organization, with the structure to hear from farmers in all parts of the country.

I was particularly impressed with the NFU leadership on issues such as intergenerational land transfers, farm diversification, rail line retention, maintaining the "Crow Freight Rate," plant breeders rights, genetically modified plants, chemical pollution, climate change, and dominant corporate monopolies. I valued their evidence-based support for the Canadian Wheat Board single-desk marketing monopoly for farmers, and their courageous exposure of how the multi-national corporations exploited farmers for industry profit. Of special interest to me was how the NFU attempted to bring farmers together with other workers in labour unions, professional associations, community associations, and governments, in order to support workers in the food production industry and promote the survival of the family farm.

The NFU has documented how billions of dollars worth of food production, is delivered into the world market by Canadian farmers and yet they only get a minor share of the revenue generated. The more powerful monopolies, duopolies, and oligopolies consisting of multi-national corporations take the lions share of the value added profit. I remember the NFU showing how the amount paid to the maker of the plastic bag, holding a loaf of bread, got more than what the farmer got for all the wheat used in the loaf of bread. A similar example showed how the cap on the beer bottle cost more than the barley used to brew the beer. Of further interest to me was how the NFU kept in touch with other farmers across the country and around the world.

One late summer day after I had joined the NFU as a new member, I decided to stop by the head office in Saskatoon. That was when I met Roy Atkinson the president and Stuart Thiessen the Chief Executive Officer. President Atkinson was a big man who appeared a little rough around the edges but with a wholesome friendliness and an obvious commitment to farmers. His office was large and full of various piles of papers and bound documents, along with several local, regional, national, and international maps on the newly painted glossy white walls. For me his reputation had preceded him as a tough guy "who took no shit" and "told it like it is," with no time for stupidity or fuzzy spin. My first encounter with Roy assured me that his reputation was valid. However, there was something refreshing about that up front, no frills, honest to goodness approach. It was also personally comforting to see he was someone like myself, somewhat full in shape, dressed in farm style blue jeans, in a stomach stretched, three-button, brown, golf shirt. Somehow a suit with a shirt and tie would not have seemed as appropriate.

Strangely, this tough, bulky, intrusive fellow had a calming smile and soothing voice when he asked me to come in and sit down. You felt like you were with someone who really cared about what he was doing and what you were doing. To my complete surprise early in our visit I discovered that he knew something about me. He informed me that he was aware of my work with the Land Bank, Farm Ownership Board and the Hall Commission. He knew that I was a single guy with a university degree, who farmed on a grain farm near Tyner in the summer, and took on various farm related employment options for the winter months.

A short way into our conversation after the small talk and discussion about the crops, Roy asked me what I was going to work at over the coming winter. I told him I had no plans. Then he said he would like me to meet Stuart who he went to find. In a few short minutes they both came back and immediately asked if I would like to work with them as their Research Officer. Without much hesitation and with some pride of accomplishment I agreed and began to look forward to my new challenge. My work would be with Stuart in the Saskatoon office as part of a total staff complement of about ten full-time people. As president, Roy would be based in the office but would be out across the country much of the time. They said that sometimes I would be on the road with him which seemed interesting to me.

On the surface Stuart was the direct opposite to Roy. Stuart was slight but tall, quiet spoken, mild mannered, academic in nature and appearance, with long gray hair making him appear like your college professor. He would try to put on a gruff arrogant boss-like demeanour, from time to time, but it didn't camouflage his extremely kind humanitarian character and his academic mind. On the issues, philosophy, and commitment he and

Roy were identical twins. I also felt a comfortable affinity for those s ideals. Therefore, my work in research with the NFU was another rew ing experience for me, working with executive board members and many interesting farmers.

A large component of my job involved communicating with the farmer membership, either at outside meetings or in my office, where farmers would visit me on a regular basis. We would have coffee, a few jokes, and some serious discussion. Each year the NFU held a national convention somewhere in Canada, which I was expected to help organize and attend. When high profile industry experts would attend as guest speakers at our convention, it was my job to entertain them upstairs at a hotel hospitality room prior to them going on stage to speak. I would meet some very interesting and important people from many walks of life. Another duty I had was to assist with some speech writing for the president and prepare him for media interviews. In that role I got in on several interesting trips with Roy so I got to know him very well and became his friend for years to come. The same could be said about Stuart.

There was one special incident on my trip with Roy and Stuart to the NFU convention in Ottawa. We were all attending our national convention held at the Chateau Laurier Hotel, just a short distance from the Parliament Buildings, on Parliament Hill. I remember it very well because I spent both business time and social time with a beautiful young lady about my age, who was a graduate student studying at Carlton University, writing her thesis on the structure and operation of the National Farmers Union. Much to my boyish pleasure it was decided that she and I should team up, in order for me to assist her with her project. Without describing in any great detail our personal feelings, I can say we quickly displayed a compelling interest in each other. It would be suffice to say that I felt like I had been sent on a heavenly mission and from her words and actions I believe she felt about the same.

At that time Hazen Argue was Minister in charge of the Canadian Wheat Board in the Trudeau Government. Mr. Argue was a Senator from Saskatchewan who was defeated by Tommy Douglas when they both ran to become the first leader of the federal New Democratic Party. After that event Argue had left the CCF-NDP to join the Liberals and later be appointed as a senator. Since the Liberals didn't have any amount of MPs from the prairies at the time, Trudeau appointed Mr. Argue a prairie senator as a cabinet minister in charge of the Winnipeg based CWB.

One afternoon at the convention, Roy asked me to phone over to Hazen, who he had known for years and ask if we could meet him for lunch. When I called we were quickly invited to come over to the parliamentary dining room the next day at noon. Argue's Executive Assistant reminded me that we needed to wear a tie to go into the dining room. I reported to Roy that the lunch meeting was set up and we agreed to meet in the lobby of our hotel the next day at 11:30 to catch a cab over to the parliament buildings. I reminded Roy about the dress code as it had been told to me. Roy as NFU President, Jim Maines as NFU Vice-President, Jean Leahy as NFU Women's President and myself as NFU Research Officer would attend with Minister Argue and three of his assistants.

When we met down at the hotel lobby, Jim, Jean and I, got there early looking very well groomed and finely dressed in our best duds, complete with suits, shirts and ties. Jean looked very classy in a lovely green dress. I looked very good in a navy suit even if I do say so myself. My new graduate student friend, whom I was quickly becoming very fond of, came with me to the lobby to see me off to the lunch, and she said I looked very handsome, which meant a lot to me coming from her. My vanity could not be hidden.

Just a couple minutes later Roy joined us at the lobby. He had obviously not complied with my warning about the dress code. I was only partially joking when I asked Roy were his shirt, tie, and jacket were. He said he was running late and forgot all about my warning. We needed to catch the cab so Roy jumped in wearing his blue jeans, with his customary brown golf shirt. Roy calmly assured me that he was sure he did not need a tie to meet his old farm friend Hazen. However the newspaper later reported that Roy was a sartorial mess.

When we arrived at the parliament buildings, Minister Argue and his group met us at the entrance to the dinning room dressed to the nines. However, also as a very large full figured man, Hazen spilled out of his shirt and pants in a messy uncomfortable fashion, which probably told us why

Roy resisted wearing formal suits. None-the-less, rules are rules when it comes to parliamentary protocol and Roy was not supposed to enter without at least a tie. In an effort to save the day, one of Argue's assistants slipped back to his office and found a tie for Roy. He then assisted our fearless leader to put on a bright red tie under the collar of his chocolate brown leisure shirt. That was indeed an emphatic farmer goes to parliament fashion statement. Together we all walked boldly through the main dinning area where I recognized Pierre Elliot Trudeau, Joe Clark and other famous faces at different tables. We were escorted past them as if we were also important, toward a very fancy private dinning room at the back. Once seated we joyously consumed a lovely meal of many cheeses, baked and mashed potatoes, chicken and steak, to be finished off with apple pie and ice cream. It was a meal fit for kings.

Speaking of kings, when I sat watching Roy and Hazen who had big statures, big minds, big egos, big causes and big hearts, sitting as friends chatting and eating with a clear interest in good food and good ideas, I thought of a movie I had seen, depicting King Henry VIII and his court. There was something special about these men that went beyond slick dress, large hair, and misleading spin. These guys were real people with their fair share of human imperfections, who had a burning commitment to help family farmers and the farm industry.

The 1980s for farmers could accurately be described as "The Dirty Eighties" when contrary to economic thought, farm production was low and prices for those crops were also devastatingly low. Off-farm employment opportunities were almost non-existent, especially government related jobs if you were not well known as a "Devine Conservative Supporter." My reputation as a farm activist and President of the Saskatchewan Crown Tenants Association, that organized and led the resistance to the Devine Government's effort to dismantle our farm support programs, did not put me in line for government employment. Right-wing politics always trumped education and experience in the recruitment policies of the Devine years. I was never on his Christmas list.

Therefore during the 1980's I decided to spend much of my non-farming winter months as a university student. I added a BA in economics in 1984, a Post Graduate Diploma (PGD) in administration in 1986 and a Masters Degree in Educational Administration in 1988 to my existing Bachelor of Education Degree from back in 1972. Sharon continued to take on banking jobs in Saskatoon for the winter months in order to complement our farm income. I again was preparing for my unknown future by enhancing my education.

However, sometime during the severe farm financial crisis of the mid-1980's, I was hired as a financial advisor by the Farm Debt Review Board. My job consisted of sitting down with farm families experiencing extreme financial difficulty to help them restructure their debt situation and deal with filing for "bankruptcy" or "qualifying for the farm debt lease back program." It was a program that offered a farm family the option to turn their land back to the Farm Credit Corporation or lending institution with the right to lease it until better times. Then they could repurchase the land.

That financial advisor experience was a heart-wrenching job, which made me care more and more for the plight of my fellow farmers. It was also quite insightful for me to see in a first hand personal context the obstacles and choices facing good people. For me it was a rude awakening on how farm finances can get away on you if your management skills are not refined to fit the shifting financial environment. I saw first hand how some folks wanted more land at any cost on the gamble that it could not lose its market value, and it would always earn enough to make the annual amortization payment. I saw how some farmers loved big brand new machinery at any cost, and thought the tax write off share of the price saved, was enough to justify the rest of the full cost. I saw farmers who beyond their means felt they needed all the toys, like trailers, fifth wheelers, quads, snowmobiles and boats along with an expensive world holiday each year.

In one case I arrived at eight in the morning to begin my work with one farm couple up in northeast Saskatchewan. We began to pour over their books with a view toward understanding what had gone wrong and what we could do about it. I must admit that things were just not adding up. Soon it was getting on to lunch time but something was still missing in our deliberations. By then we all felt like we knew each other quite well so the farm couple asked me to join them for a lovely farm lunch.

After lunch as we got back to the financials, the couple almost in unison said to me that there was something they had not yet shared with me. I had thought so. Unfortunately this couple had a son who was almost continuously in serious trouble with the police and the law. Simply put, this irresponsible son had been draining this farm family of its precious finances due to lawyer fees, penalties and living costs when he was not in jail. With his parents' reluctant agreement he had swamped the farm financially.

Perhaps I should have had more sociology and psychology courses and more related experience under my belt, but I could only speak from the financial documents and from my heart. I told them of the accepted practice to pay yourself first and your business next before any other costs. You need to be able to live comfortable but frugal and take care of the farm bills in order to survive in the farm industry. I also said quite sharply that if you folks are ever "broke financially" then you can be of no financial assistance to any of your children and most of all you will not get the rewards you deserve from your own work. At the same time your confidence, dignity and happiness will be at risk. You owe it to yourselves to recognize these realities. Some bold advice I suppose.

In a soul searching private moment inside that clean bright kitchen, surrounded by the soothing aroma of home baked bread, while we were sitting at a large golden oak table, in a touching reflection of common humanity, our six hands came together farmer to farmer to farmer. A warm and radiant smile from the farm mom seemed to silently promise me that we would work together to solve this farm debt problem. The dad with a little bit of a reluctant tear on his weather worn cheek, told me that he knew what I meant, and agreed we should go forward with my plan for their farm survival. That we did and it saved their family farm.

Later as I drove home that afternoon I told myself out loud in the confines of my own truck, how important it is for farmers to work together to support each other. Disposing of our neighbours, our friends and our communities is not right. Sending farmers off the farms they love into the city they fear, to take jobs away from struggling city folks, is not a meaningful solution to the unfortunate problems of the farm financial crisis.

Perhaps my greatest personal employment contradiction at that time in my career was that I held so many education related credentials but had not yet taught full time in the classroom. Yes I had done my practice teaching requirements but I had never taken a formal teaching job. Teaching for me had not been a priority over my other government job opportunities. However, in 1989 after I completed my Masters Degree I started to think about going teaching, if I could find a job, and if our farm could be otherwise looked after.

Prior to the 1990–1991 school year, I saw an ad in the paper saying that the Meadow Lake school division was looking to hire a teacher, to design and implement a new Canada Jobs Strategy-Work Experience Program, at Ernie Studer School in Loon Lake, Saskatchewan. The ad said to contact Bob Lockwood the Assistant Director of Education. Bob had been a classmate of mine at university when I did my masters degree. I called Bob who said he was happy to hear from me. Then he invited me to meet him and the Local School Board at Ernie Studer School within a few days. It was the first time Sharon or I had ever been to Loon Lake but within one hour of our arrival I had accepted a teaching job offer.

That venture from the farm to Loon Lake was a big event for Sharon and me and our boys Danton age-eight and Brandon age-seven. However putting it into perspective, it wasn't the first time that farmers on the prairie had, for financial reasons, moved north to the Meadow Lake region. I had heard many stories of how folks during "The Dirty Thirties" loaded all their worldly possessions onto wagons driven by horses or vintage vehicles heading north to the Meadow Lake area. Our move was not quite that dramatic but it certainly did uproot our established lifestyle. It was necessary for us to hire someone to work with my dad on our farm. We were able to find another struggling local farmer as our very dedicated and capable hired man. As a result my family was able to move to Loon Lake, a small town of about one hundred and fifty people, located right beside Makwa Lake and an Indian Reserve, forty minutes west of Meadow Lake.

My job depended on good community relations because I needed the goodwill of local businesses to have my work-experience students placed at all kinds of jobs in the community. Therefore I got to know the local people very well. Unfortunately my dad became ill so my family and I returned to Saskatoon after only the one school year in order for Sharon and I to again commute to the farm.

Once back in Saskatoon, but commuting to the farm, in the summer of 1992 I got a call from Harley Olsen, now a senior manager with the provincial government, still located in Saskatoon. He asked me to drop by his office at Innovation Place whenever I was back in the city. I had known Harley for some time as a friend and as a colleague. He had been an usher when Sharon and I got married in 1980. We had become friends early in the nineteen-seventies when we both worked with the provincial department of agriculture in Regina. When I was Acting Director of the Saskatchewan Farm Ownership Board I had hired Harley as research officer.

Harley Olsen a life-long civil servant and political survivor had turned up as head of (PARD) the Federal-Provincial Partnership on Rural Development. When we got together over at his office he described to me the objectives of the PARD program. It would be administered by a partnership consisting of employees of the federal government in the (PFRA) Prairie Farm Rehabilitation Agency and provincial government employees in the Department of Economic and Co-operative Development. Potential economic development projects would be evaluated and if approved partially funded by PARD to do feasibility studies, business plans, advertisements, prototype development or human resource development. The overriding objective was to create new business activity in Saskatchewan.

Harley said that he wanted to hire a Rural Development Officer to work directly with him setting up the PARD program and explaining it to the public and potential clients. That person would be assisting with the application process, participating in the evaluation and selection procedures, as well as serving as the project manager to insure that funding was utilized as intended. He said he wanted to offer me that job as a full-time position for a three-year period. After three years the position by labour rules would need to be classified with the union and made available to a union employee with more seniority than what I would have after only three years. Harley said that given my university education, my previous experience setting up government programs, and our previous positive working relationship, he could get approval to hire me for the three years. I gladly accepted even though it meant my wife Sharon and I would both be working full-time in Saskatoon, while also doing our farm work on weekends and holidays. A "Herculean Task."

In order to facilitate my return to government and because the final arrangements for PARD were not yet made, Premier Romanow offered me a short term three to four month contract to serve directly as his Ministerial Assistant located at the Cabinet Office in Saskatoon. My first assignment there was to co-ordinate the Sask. Power security systems necessary to secure and clean up the old A.L. Cole Power Station site near the river in downtown Saskatoon. Then I was given the task of working with the Saskatchewan Water Corporation, who controlled the water level in the South Saskatchewan River by changing the water levels at the Outlook Dam. It was a high water time of year and the "Power Boat Races" were scheduled in Saskatoon, but they were in danger do to high, fast and dangerous flows of water and floating obstacles like tree trunks and drift wood. I recall that we couldn't back the flow up at the dam for very long because it would flood farm irrigation pumps. However if we raised it a reasonable amount at exactly thirteen hours before the races were to begin, the beneficial effect would arrive in Saskatoon at the right time and the dam level could then be quickly returned to normal. Our plan was a success. What I remember most about my work with the Premier was when you called and said you needed something done, it became top priority and it happened much quicker than I even expected. All you had to say was: "This is Dan Danielson from the Premier's Office and I need your help with something." The response would be positive.

One day I remember the Premier said to me that he wanted to introduce me to Rusty a devoted constituent in his Riversdale Constituency. He said Rusty is a great guy so he was hoping I would entertain him when he frequently came into our office. To this day Rusty and I are great friends and I often remind him how I used to get paid to drink coffee and tell jokes with him. "Orders from the Premier."

The two most often asked questions I got when I worked in the Premier's office were. Does Romanow have a girl friend on the side? And do you think Romanow is going to jump ship and join the Liberals? My honest answer always was no for both inquiries. In my humble opinion the idea that he was destined to join the Liberals was mainly wishful thinking on the part of the local Liberals. I also know that he always kept in close contact with his wife.

Moving over to Innovation Place at the University of Saskatchewan to join Mr. Olsen and the PARD team was a great pleasure for me. Our office space was modern, large and almost new with all the technology, and a very well qualified administrative staff to work with us. We had a beautiful boardroom that served as a meeting space and coffee room. The economic and co-operative development team I worked with included specialists in Information Technology, Biotechnology, Construction, Engineering and Tourism. My focus on the team was as a Rural Development Specialist. Some of the business projects I remember working on included an ethanol plant, a plant biotechnology company, a malt company, a salad processing plant, a pasta plant, a food catering business, a cattle feed lot, a golf coarse, as well as the commercialization of newly invented prototype products.

During the last part of my assignment with PARD I attended an international conference in Saskatoon called "Economic Development and Co-operatives Into the 21st Century." More than 500 delegates from 22 countries representing large and small co-operatives from virtually every economic sector attended. All the participants had a common commitment to co-operatives that drew them to Saskatoon. The conference focused on the following sub-themes:

1. Co-operatives in a Changing Global Environment.
2. Co-operatives and Community Development: Responding to change.
3. Co-operative Structures, Principles, and Business Practices: New approaches and models for co-operatives in the 21st century.

Garf Stevenson a former president of the Saskatchewan Wheat Pool served as conference chair. After the conference Mr. Stevenson made this statement at the release of the "Report on the Conference Proceedings," which I had written.

Mr. Stevenson said: "Given the numbers, and diversity in backgrounds of the participants, and the diverse country of origin of those who attended, the quality and variety of material, which was presented, the decision was made by the Co-operative Directorate of Saskatchewan in the Department of Economic and Co-operative Development to prepare a summary of proceedings, as a permanent record… This was a substantial undertaking and it was decided to engage someone who had participated in the conference, possessed the academic credentials and experience and had a good knowledge of issues facing co-operatives, to undertake the project. Dan Danielson, a farmer and economist with graduate training in education and considerable experience with co-operatives and agricultural issues, agreed to undertake the task. The result is a document that summarizes the eighty presentations and related discussions and stands as a valuable record of that important event." My work renewed my commitment to co-ops.

seven

ADVICE ON BOARD AND COMMITTEE GOVERNANCE

One time I asked my professor in a policy governance class the following question for clarification: "Did you say a board or committee is usually made up of an odd number of people sitting around a table making decisions, or did you say that usually there is a number of odd people sitting at the board or committee decision-making table, or did you say a number of bored people sit around an odd shaped table making decisions at board or committee meetings?" "All of the above," he quickly replied. Then he went on to say that governance is all about leadership, and is rooted in concepts like trust, commitment, intelligence, vision, humility, respect, empathy, patience, integrity, courage, perseverance, inclusiveness, healthy-conflict, caring, generosity, and charisma. Leadership in his view is not so much about the power of a position in authority as it is about the personal characteristics necessary to influence others. Leaders are influencers he would say. In his opinion, developing, maintaining and refining those personal characteristics, along with showing the collective ability to produce meaningful results, is the definition of leadership. In action terms, he suggested that effective leadership means listening, planning, directing, delegating, coaching, supporting, participating, communicating and selling. It is the process of organizational problem solving, not playing personal or political games. Leadership means service.

I have become absolutely convinced that more and more of the important decisions affecting our everyday lives are being made by fewer and fewer people in elite positions of power and influence. The general public, I fear, has drifted into a dangerous culture of indifference and apathy, often generated by a feeling of helplessness. It is like they are sleep walking into a complex future. The reality is that if people don't contribute or participate, by default they surrender their democratic right to others often less qualified than they are. The irony is that there are also hundreds of folks all around us going to thousands of meetings who are unwilling or unable to express their views or produce the results they desire because of dysfunctional organizational structures, processes and participants. I believe that improving our board and committee governance environment is essential to restoring the hope and confidence required to recruit and retain active new leaders. Our democracy depends on it.

Thinking back to my boyhood at Tyner, I guess my first crack at leadership came as a member of the local "Boy Scouts Troop." I was chosen by my youthful troopmates as a "Patrol Leader" for the "Eagle Patrol." To this day, I recall my pride to have been selected by my friends and supported by our dedicated leaders. It was a job I took very seriously. Even if it meant "sleeping" out at "Poplar Bluff" on sharp roots and cold hard ground, in a thin green sleeping bag that by morning had flipped, flopped and twisted into the shape of a bow tie. Later on as a very young bachelor resident of Tyner I remember serving on the "hamlet board" for the town. Better yet; I was designated by my fellow board members as the "mayor"—a title I would have some fun with to this very day. As I recall even by then our town was losing its population, so when folks moved away it was our obligation as the town leaders to move their street lights to a new location of greater utility.

Over the years I have held several elected or appointed positions in the field of board and committee governance. I've served on three Saskatchewan Wheat Pool committees, and as President of the Saskatchewan Crown Land Tenants Association. I have been elected five times and continue to serve on the Saskatoon Public School Board as trustee and previous board chair. Our school division is the largest in the province. I was elected several times as a Provincial Executive Member of the Saskatchewan School

Boards Association (SSBA), and the SSBA then appointed me to represent education as a Provincial Board Member for the Saskatchewan Assessment Management Agency. More recently, I have been elected by Saskatchewan Wheat Producers as a Director and Vice-Chair of the Saskatchewan Wheat Development Commission. I have also served as a corporate director with the Primrose Resources Economic Development Corporation in northwest Saskatchewan. In addition to those board positions, I have been elected or appointed to various governance committees such as the audit committee, finance committee, pension committee, hiring committee, risk management committee, HR committee, and communication committee.

The 2005 SAMA Board of Directors (left to right): Sharon Armstrong, Ron Thomas, Rick Brunsdon, Fred Clipsham, Craig Melvin, Arthur Scherr, Robert Schultz, Darryl Senecal, Dan Danielson, John Wagner (missing: Richard Douglas)

On my personal journey both at the farm and in the city, I have experienced some rewarding successes and also observed some very questionable and dysfunctional organizational behaviour. One of my most rewarding successes as a trustee with the Saskatoon Public School Division was the transformation of an underutilized elementary school into Royal West

Campus, operated as a "Second Chance High School." This new associate campus would offer flexible programs, flexible timetables, and special teachers and administrators to assist students who had fallen out of the regular high school programs. These would be students with ability who had met various obstacles such as pregnancy, illness, or family tragedy that stopped their academic progress. Now at an older age than the mainstream students in the same grade, they can rejoin the school system and get back on track to high school graduation in a more responsive environment.

There was growing pressure on our board to close the school and sell the building. We would save about ninety-five thousand dollars per year. However, because it was in my Ward 4, as trustee I was assigned the task of coming up with a recommendation for a board decision. As a result, I contacted Principal Brian Hilson at Mount Royal Collegiate to seek out his advice. Together, we soon came to the conclusion that there was a need for a "Second Chance School", and that Brian and other administrators would come up with a documented plan. Within a few weeks, I had a comprehensive plan approved by the board. We knew at the time that if we could get sixty-nine or more students to participate in the program, it would be a financial success. We got that number and a few more the first year. At last count, in this twelfth year of operation our enrolment at Royal West Campus is nearly four hundred students.

Each spring at graduation time it is absolutely heartwarming for me to speak at the graduation ceremony and congratulate the graduates and their families. I always tell the story of why we wanted this Second Chance School. Incidentally, our school division was awarded the Premier's Award for Innovation and Educational Excellence for our Royal West Campus. I was proud to accept the award as chair of our board and give a speech in Regina explaining why and how it came about.

Later, as Chair of the Saskatoon Public School Board, I had the privilege to officially turn the sod and oversee the construction of two new high schools—Centennial Collegiate on the east side of Saskatoon, attached to a state-of-the-art indoor soccer centre, and Tommy Douglas Collegiate on the west side of Saskatoon, attached to a new state-of-the-art, Olympic class aquatic centre. Each of those new public schools proudly boasts a four

Royal West—a "Second Chance High School"

Tommy Douglas Collegiate and Shaw Centre

hundred and fifty seat theatre complete with orchestra pit and all the necessary stage production amenities. A new high school and community theatre on each side of the city—how much better could it be? I was happy to do my part.

From family farm to city boardroom the country stays in the boy. I was very proud to be able to introduce a motion at our board table to begin the establishment of community gardens at eight of our elementary schools with a view toward expanding that program throughout our city schools. Thanks to my fellow trustees, that motion received unanimous approval.

Much to my delight, in 2011 my colleagues on the Saskatoon Public Board of Education unanimously nominated me for the Saskatchewan School Boards Association Provincial Award of Distinction. Their motion stated: *"Dan has earned the respect of his fellow trustees for fairness of his approach to managing meetings, insightful leadership and a wealth of knowledge he shares so freely. Dan has proven his ability to lead in times of change, to advocate for local autonomy in education, and to ensure trustees are not distracted from their critical governance responsibilities. Since elected, he has provided outstanding service to public education in Saskatoon and throughout the province."*

Then on Nov 12, 2013 I was fortunate enough to be awarded a "Lifetime Membership" in the Saskatchewan School Boards Association (SSBA), based on a contribution formula representing years of service and participation on a certain number and variety of boards or committees. My acceptance speech at the SSBA annual general meeting banquet allowed me an opportunity to express some concerns I hold about how school board autonomy, both in areas of finance and programming, has drifted into the centralized control of the ministry and the government. The K–12 public school system, in my opinion, has become highly politicized in the spirit of "LEAN" a cost cutting vehicle thinly disguised as an efficiency doctrine.

Here is the substance of my presentation:

> *"Good evening fellow educators. Since I turned sixty I have felt compelled to offer free advice every time I get a chance. My family in the audience will probably tell you that I started this crusade a while before now. Nonetheless, I know it is common practice on these occasions to provide a few observations. Therefore, I want to make some remarks with respect to*

SSBA Lifetime Membership from
President Janet Foord

the future relevance of the SSBA and the local autonomy of the twenty-nine school divisions it represents. Over my years as a trustee or board chair that autonomy issue has come up quite regularly.

I believe the SSBA can have a relevant and productive future because there are always goals or objectives we can accomplish together that we can't do alone. However, an effective SSBA going forward will depend on the courage and competence of you as trustees, executive members and staff. You will need to be strong leaders, not "leadersheep." If the SSBA is not willing to develop and express strong evidence-based positions on major education issues, it will become irrelevant and vanish. You owe it to yourselves, and more importantly you owe it to our students, parents and taxpayers, to put down your rubber stamps and raise your collective voices. The SSBA, in my view, should not be a mutual admiration society, a social club or subordinate division of the provincial government. You need to remember who elected you, and whom you are expected to represent. You are actually assisting the government and the public the most, whether they recognize it or not, when you come forward with strong justified positions for improving education, even if that position is not the accepted position of the government ideology. I didn't actually hear him say it, but I think it was Einstein who said "When we all think alike, we all stop thinking."

If anybody tells you the province now pays all the cost of education because they eliminated your local authority to set the property tax rate, you need to say it is not true. Even though your local board no longer sets the rate of taxation (mill rate), local property taxpayers still directly pay thirty-five to fifty percent of the cost to fund our schools, and they have elected you to represent that investment. Those same local property taxpayers also pay provincial taxes and fees, which help fund the remaining provincial share for construction, maintenance and operation of our schools. The people you represent have all of the skin in the game.

If anybody tells you that you cannot be part of the decision making process, but you need to implement government or ministry decisions, then you need to jump up and defend your "Board Autonomy." You must insist that the responsibility to implement major decisions, without the authority to help make the decision, is not good enough—especially if those decisions are made off-the-cuff and on-the-fly for mainly short-term political reasons.

If anybody tells you that full-day-k is just a babysitting service we can't afford, or that standardized testing is a higher priority than improved instruction, or that local agreements with teachers are too lucrative, or that teachers should not get a fair salary increase, or that our school systems are not LEAN enough, you should defend our students and our teachers by taking a strong position based on empirical evidence, academic research and good management procedures. Our teachers and our employees are not our enemies and ***"Teaching is the profession that produces all other professions."***

If anybody tells you that publicly funded school boards can't build state-of-the-art schools, on time and on budget, and that the private sector needs to step in and bail you out to get the job done, then you need to stand up and ask some important questions. Don't they know that the Saskatoon Public School Board built two twenty-six million dollar high schools, one attached to a soccer centre and another attached to an aquatic centre, on time and under budget within a two and one-half year time period? (I know, because as board chair I turned the sod and helped guide those developments. Each are modern schools with a 450-seat theatre and orchestra pit.)

You must respectfully ask these P3 Public-Private Partnership proponents what the private partners bring to the table. Will the private sector who also benefit from schools actually bring cash grants to the partnership? Will the city, which benefits from the school, bring the land at no or low cost to the partnership? Will the construction price for the school actually be lower? If so why and how much lower? Will the interest rate for the funding be lower than government rates? How much will an intended longer amortization period increase the overall cost? Will the bidding process be open and fair, and will local suppliers and workers be part of the process? Will the school and public stakeholders' access to the new facilities be restricted? Who will carry the debt on their books? Are these P3 arrangements back-end loaded with many of the costs kicked down the road for future generations to pay?

You must not fall blindly for the religion of private enterprise, which says "Have faith and pay us more because private enterprise is always better." Find out how it will be better or if it will be better. Remember that any wise business includes a healthy profit margin above the real costs, which in the P3 world they mysteriously call a "risk premium." As boards we need to know what we are getting into.

There is no doubt that we need many new schools and we need them yesterday. However, my friends, even in our state of severe desperation we still need to do what is accountable and responsible for those we represent. I know you are too vigilant to fall for that age-old strategy, where a crisis situation is used to convince otherwise thoughtful folks to unwittingly accept an ideology-driven one-sided solution. Remember that we are the "Public School System." Public ownership and control is our mandate. The people in our province are counting on you to do what is right, not what sounds right, embellished by the "spin" of political expediency. I know you will make the right decision. You can do it. Our future depends on you and our students will thank you with their success.

Thanks very much for this evening, and this recognition. I really do appreciate it."

The demise of the Saskatchewan Wheat Pool has been a difficult experience for me to accept, having been involved with the Pool for many years and having served as a "Local Committee Member and Chair." It troubles me to accept how a failure in governance and a resulting lapse in member participation resulted in such a terrible loss to farmers.

On April 2, 1996 Saskatchewan Wheat Pool shares began trading on the Toronto Stock Exchange. The share conversion required a change in provincial legislation, which was passed by the Romanow government. Initially, there were two classes of shares offered, one to limit ownership to farmers and the other class for investment only. The initial shares opened at $12, rising to over $24 in 1998, then down to $9 in 1999 and $.20 in 2003. In 2005 the two classes of shares were consolidated and the ownership limits were removed, transforming the SWP from a co-operative to a privatized business corporation. In 1999 the SWP posted its first net loss of over $14 million and in 2000 that loss went up to just under $100 million. The share of grain deliveries it handled also fell from the constant sixty percent of market share down to thirty-five percent.

In a November 2006 paper by Murray Fulton and Kathy Lang from the Department of Agricultural Economics ("The Saskatchewan Wheat Pool Share Conversion," page 7) they say: "There is considerable evidence that a principal-agent problem (board-management problem) was at work in the SWP. The share conversion marked a significant shift in decision-making power from the Board (principal) to Senior Management (agent). The conversion dramatically changed the skill set required by board members. One of the commonly expressed views was that the board did not have the expertise to run a publicly traded co-operative. In addition, the share conversion itself can be seen as a vehicle introduced by management to both change the corporate culture at the Pool and speed up the decision making process. One of the conclusions of the analysis is that the share conversion acted to sever the economic linkage between the Pool and its membership; at the same time, the Pool's governance did not ultimately generate adequate investor confidence."

Then on page 15 the paper said: "A shift in power from the board to the CEO created a second impact. With the CEO and the president, there was the quick transformation of power from a president's office to a CEO's office. There was a sense that "management was running astray with the company." Under the CEO between 1996 and 1999 the SWP directed funds to twenty-five new or expanded ventures worth hundreds of millions of dollars, to build high throughput elevators and to invest in Humboldt Flour Mills and terminals in Europe and Mexico, regardless of expected profitability. By 1999 the CEO was fired from his position.

Fulton and Lang say on page 17: "The board seemed to be missing the individual who was just a cynic and not afraid to ask the simple hard questions or the critic who would force management to think it through and come better prepared with proposals. Part of this lack of ability to serve as the final check is attributed to the process of board elections. Increased board appointments by acclamation had a major impact on the types of people who were put forward as board members. The idea that the quality of board member falls as competition for positions decreases was definitely evident in this case." The board of directors had failed to do its job and the Saskatchewan Wheat Pool was soon history.

At the time when the Pool was privatized, Sharon and I had seventeen-thousand five-hundred dollars earned and held as equity in the company. We decided to take ten-thousand dollars out and leave seventy-five hundred dollars in as our share value. With the demise of the Saskatchewan Wheat Pool those shares became worthless.

In another governance situation I was elected as chair of "The Committee For A Democratic Co-op" which has come together in Saskatoon to deal with an issue of failed democratic governance within our co-op. There are approximately one hundred thousand members of the co-op, but the voting process to elect the board of directors and vote on policy resolutions, is restricted by the requirement for members to physically attend the annual general meeting (AGM). Only about three hundred members come out. It is a restricted democratic process reliant on less than one-half of one percent of the members for its democratic control. To complicate matters, the management wing of Federated Co-operative Limited, which is contracted to handle the management functions of the co-op, is suspected of packing the AGM to manipulate the democratic process. A portion of the membership say that as long as a good profit is made, we should be willing to give up the democratic structure set out in the basic co-operative principles. That would eliminate our co-operative advantage as we become just like any other store. The result of this negative circumstance is that members including some previous board members have been marginalized out of the democratic structure. Therefore, our committee has mobilized to work for a more responsive and inclusive decision making process consistent with the co-operative principles. More specifically, our committee has developed and presented these recommendations for improving the Saskatoon Co-op governance structure:

1. **Democratic Control by Equal Voting Rights and One-Member= One-Vote.** Currently only about 300 members out of 100,000 total members are able to attend and vote at the AGM (less than ½ of one percent). No facility can accommodate them all. *Therefore, we recommend expanding the voting opportunity to the various business locations.* The 1996 Co-op Act enables changes in the Co-op by-laws to make this change.

2. **Autonomy and Independence with Democratic Control.** There is a real concern among our membership that total control of our co-op, including both the policy and management functions, has been transferred to Federated Co-operatives Limited. *Therefore, we recommend that the current management agreement between Saskatoon Co-op and FCL be reviewed and renegotiated to secure membership control of the policy—making function and restrict FCL to a delegated management function.*

3. **Education, Communication and Member Relations.** Quite frankly, our co-op membership does not feel that it is being consulted with or adequately informed by our board of directors on most policy matters. Some feel that the Co-op board has thrown up its hands in frustration and allowed FCL (the wholesale management wing) to make all policy decisions. Many feel they are only notified of changes in co-op policy

after the fact, with no input into the decision. When information and input around policy decisions is requested, the board often takes on a defensive bunker mentality, as if the members are the enemy and we only get in the way of the accepted decision making process. *Therefore, we recommend that our elected Co-op Board, in its policy role, should adopt a new consultation and two-way communication policy, with the membership. Resources should be allocated to engage Board Members in proactive dialogue with the membership. Published and distributed minutes from meetings with recorded votes should be adopted. A qualified and committed education and member relations employee, with the required resources to succeed, should be assigned to implement and manage a comprehensive policy consultation and member relations program.*

In my lifetime I have witnessed the demise of some rural co-ops, the disappearance of the Saskatchewan Wheat Pool, the virtual elimination of our Canadian Wheat Board, the marginalization of the membership in the Saskatoon Co-operative Association and other existing co-ops, along with the erosion of local autonomy away from school divisions. I have observed how governing boards have allowed their fiduciary trust commitment with the membership and owners to drift away or be taken out of their hands.

Over my years of service I have discovered that one person alone never is totally the reason why accomplishments are made or failures happen, because the power for success or failure is almost always a collective challenge. Success or failure is often found at the end of a long messy and time-consuming trail of meetings and more meetings. As a result, the development of caring human relationships and positive organizational cultures is extremely important to generate and sustain effective decision-making processes. Those partnerships and processes more generally applied, in my view, are essential for the advancement of our modern democratic society. So what can be done to improve our governance structures, processes and participants? On that topic let's take a moment to review some relevant organizational theory as it is generally described in the literature. I have lost count of how many books or articles I have read or seminars I have attended on this subject.

Simply put, board and committee governance is required whenever people come together from the private, public or co-operative sectors with a common purpose and a desire to get something done. In more academic language this process is referred to as "Agency Theory." On the one side of this organizational structure you have the owners, shareholders, members, electorate etc. known as the "Principals", who usually share similar values, beliefs or concerns and generally know "what they want done." However they need the expert advice and assistance from people who know "how to get it done." That side of the organizational structure includes the "CEO" and his/her management team or staff who are all known as the "Agents." They are the hired guns. (*See figure 1.*)

In order for this "Principal-Agent Relationship" to be productive and produce the expected results, effective leaders need to come forward. The "Principal" delegates it's authority to a "Board" through some type of elected or appointed selection process. The board is usually established by legislative authority in an act or regulations to become a small workable group trusted with the fiduciary duty to do what is best for the organization. The enabling legislation sets out the authority boards have to develop governing by-laws and policy. Members of the board and the sub-committees they form become the leadership for the principals. A chair of the board without any official powers beyond other board members is selected by the board as its leader and spokesperson on policy matters. Together, they clearly direct management and protect the owners.

A major function of the board is to hire and delegate the administrative function to the "Agent" or CEO, who in turn hires a management team and staff to carry out the business of the organization. The board has a duty to direct and protect the CEO, management and staff, and to evaluate their performance and results. Within the board structure, the chair is the communication vehicle to converse directly with the CEO, who is the leader and spokesperson for the administrative team.

Together the board and management in a transparent, trusting, and open communication culture, regularly conduct the "strategic planning process." That process works to identify and document the organization's vision, mission, values, goals, and objectives. As part of that process, the

Fig. 1: **Agency Theory**

delegation *delegation*

principals ---------- **boards** ---------- **agents**

trust (boards mediate between principals and agents) *trust*

- owners
- shareholders
- members
- board selection

- committees

- CEO
- management

Strategic Planning
(what to do)
- strategic plan
- budget control
- compliance

Operational Plan
(how to do it)
- work plan

(boards with management establish values, vision, mission, goals, objectives)

leaders from both the board and management conduct an "environmental scan" to understand the externalities that may impact on the organization. In addition they explore the internal characteristics of the organization in terms of a (SWOT) analysis of strengths, weaknesses, opportunities and threats. Based on that analysis, the board on behalf of the owners, allocates the required resources in a budget. Then, the plan is given to management for implementation. Every person in the organization needs to be clearly informed of the board's "Strategic Plan" and the CEO with his/her management team is called on to design a comprehensive work plan to implement the policies set out by the board.

Important decisions can be made by majority vote but unanimous decisions are highly desirable. Boards are expected to establish an annual "work plan" which sets out what they expect themselves to be doing during the year. Some of a board's more specific duties include adopting rules and procedures for meetings, reviewing the "board package" and preparing for meetings, adopting meeting agendas, reviewing and adopting minutes from previous meetings, holding regular strategic planning sessions, holding regular policy reviews, adopting regular results evaluation procedures, conducting annual CEO and board performance reviews, developing an annual report, and engaging legal council and external auditors to ensure compliance with laws and obligations. Most large organizations adopt a dual audit structure, with an internal audit process to monitor and account for resource allocations inside the organization, and an external audit policy to account for resources entering and exiting the organization involving outside agencies. Developing and nurturing working partnerships with all other agency stakeholders is an important role of the board as it engages with the larger community.

The literature concludes that policy boards should delegate the following duties to the CEO and management:

- provide a clear vision, direction and focus for implementing the organization's strategic plan and establish comprehensive reporting vehicles such as regular updates, or quarterly and annual reports
- establish services, programs and activities within the established budget
- provide complete and timely information with options and recommendations to guide board decisions
- decide on the organizational chart and staffing requirements by designing staff jobs, instructing staff members, rendering any judgments or assessments of staff performance, establishing positive and professional working relationships, providing effective educational leadership and professional development and by determining staff development needs, promotions and terminations
- empower everyone in the organization and external stakeholders by transparency, accountability, delegation and consultation

Further the literature seems to agree that the most common board types are:

- **Operational Boards**—where board members get involved and actually perform the functions of the organization as well as make the decisions.
- **Collaborative Boards**—where there is very little board independence in a high trust environment where the board and CEO work together but the CEO is given most of the decision making power.

- **Advisory Boards**—where the board only sees its role as supporting the CEO and the board and principal owners are marginalized outside the decision making loop. (Sometimes called a rubber stamp board)
- **Policy Governance Boards**—where the board exerts its independence and delegates certain defined authority to the CEO and management, in order to accomplish the vision, values, mission, goals and objectives set out in the strategic planning process and supported by the allotted resources.

For most boards they will display some combination of the characteristics from more than one of these board types. Thus the flexibility to adopt a governance style from a variety of governance models is important.

Some generally accepted functions of good governance are:

- protecting the ownership and promoting key stakeholder involvement
- focusing on the needs of clients or consumers
- nurturing a positive relationship with key stakeholders
- maintaining a strong focus on results
- developing a process for monitoring achievement
- enhancing competent board and staff leadership
- maintaining a cohesive leadership team
- directing management with clear lines of accountability
- maintaining board independence (from management) when making decisions
- ensuring the effective use of resources
- maintaining financial stability
- maintaining a culture that encourages teamwork
- sustaining a commitment to board self-evaluation and development
- encouraging a healthy balance of stability and innovation
- cultivating healthy conflict
- building internal and external trust
- speaking with one voice when policy decisions have been made
- allowing members to express independent views before a decision is made

- saying no to compromise when a more complete result can still be gained
- encouraging new leadership and developing succession plans

The policy board holds the responsibility and liability for all actions of the organization and therefore needs access to all the information. Cooperation between board and management is of special relevance in the area of risk-management by helping the organization to see gathering storms and navigate through them. However for the board to take action it must follow the prescribed collective protocols. Therefore as board members individually it is a situation of "nose in-fingers out."

Some obstacles to good governance are:

- an organizational history which resists change
- a lack of knowledge on good governance practices
- lack of time, human, and financial resources
- other priorities that are seen as more pressing
- conflicting personal or political agendas of board members
- a board chair or CEO who resist board control
- conflict between board members or the board and CEO
- labor and management conflict

Over the years I have come to recognize different types of "board member styles." I think there are two main categories. On the one side of the coin there is the "results oriented group" who are a take-charge type of people, who wade in on almost every issue in the context of what needs to be done now. They are calculated and devoted decision makers, with the confident attitude that as a collective, they can walk on water. As a result, they do what they feel is necessary to get the results they want, and they are prepared to let the political chips fall where they may. This board member style is committed to be a solid "leadership team." They are not afraid of their own shadows and believe that boards need to take positions based on the information they have and what is best for the organization they represent. They are prepared to resist the pressure from senior governments or private interest groups or even the general public in order to make what they consider to be the correct decision for their organization. They are also ready to resist the influence of their own CEO when they feel it is necessary. Often these board members are strong proponents of "local autonomy" for boards. They say they are closer to the action at the local level and the advice they get from their management and staff is more directly obtained than information used in more centralized decision-making. It is their view that some decisions do need to be made at the central or senior levels, but that process is better served with assistance from effective consultation processes at all levels in the organizational structure.

On the other side of the coin there is the "like electricity group", who always take the path of least resistance and remain careful that they don't rock the boat or ruffle any feathers, with the hope that by tip toeing over egg shells, they will be able to get re-elected. As part of this "leadersheep team" you will often find the "cheerleader" cake and tea party board member. These people spend much, of their time attending social functions, drinking tea or coffee and eating cake, pie, or cookies. They love schmoozing and social climbing, which is the main reason they came on the board in the first place. However, when it comes time to prepare and deliver on tough decisions, on serious financial or human resource issues, they fade to the background and have little or no contribution to make.

Unfortunately, sometimes these are folks without much background, commitment or preparation who may have joined the board for the wrong reasons. This group is highly susceptible to corporate, government, or interest group influences, contrary to the true objectives of the organization they are elected or appointed to protect. Usually these folks arrive after a hollow popularity contest without the independence necessary to question and resist harmful external influences. Often, appearances and personality alone are allowed to substitute for the necessary substance. There is a real danger that these members will become too cozy with management or external groups, moving into the area of "rubber stamping" management recommendations or promoting outside causes, without proper analysis. From time to time they are even planted in an organization to champion issues for a senior government, a corporation, or a political party. At that point, a board without independent analytical skills and valid information can digress into a provider of unreliable decisions, supported by unbalanced "spin cycles." They become apologists for some interest group.

There are a number of practical tips or hints I have learned the hard way over the years that might be helpful:

1. Traditionally, much of the analysis and debate on the subject of agency theory or the governance and management relationship has to do with the degree of independence between the "board policy function" and the "management administration function." Sometimes the analogy of sandboxes gets used to explain that physical working arrangement. The metaphor implies that each group or side, board or management, have a distinct environment or culture, their sandbox, in which they exist, allowing the independence it requires to get the job done. If governance represented by the board interferes too much in management affairs, and makes day-by-day operational decisions outside its own sand box, it is said to be "micromanaging." If the governance board is too reluctant to make policy decisions, and allows management to make their decisions for them, they are said to be "rubber stamping" management decisions. In my opinion, there is a practical trust position somewhere in the balance that works best, because both the board and management need to be engaged in their own way with all decisions, either directly or indirectly, since they are all responsible for the results. Returning to our metaphor, the sand in both sandboxes belongs to everyone in the organization. There must be a pragmatic, workable arrangement, reinforced by transparency, open communication and trust. The glue that holds that ideal trust relationship together is the super-ordinate goal of organizational success, shared by the board members, management, staff, customers and all stakeholders connected to the organization. Success comes down to the commitment of all the people in all of the sandboxes.

2. Hold tight on all aspects of your local or regional autonomy. The loss of autonomy for the board as it is centralized to a senior government

or corporation reduces board effectiveness and increases the potential for the unhealthy politicalization of board members. Their role is forced to shift more to advocacy and influencing rather than decision-making. What can be very restrictive and dysfunctional for boards is when senior government or corporative power reduces or eliminates the "board authority" but relies on "board responsibility" to use their reputation and proximity to the situation in order to sell a position they did not make or support. Relevant board authority and responsibility held securely within the control of the local boards results in greater engagement, participation and effectiveness.

3. The selection of the Chair of your board and the hiring of your CEO are two of the most important decisions any board makes to represent its organization. There is great harm that can be done if your chair or CEO is not a good fit, either personally or professionally, for your organization. The Chair and CEO will also need to work well together in all areas of board governance, including setting the agenda for meetings. If either the chair or CEO are controlled or excessively influenced by people or forces outside the organization, there is trouble waiting to explode. More than anyone else in the organizational culture, those two positions hold the highest moral and legal imperative as servant-leaders to guide and represent your organization. They become the face and the voice for your board and the entire organization. A first clue that real danger is lurking in the wings is if one of those folks begins to handle in house board matters at private coffee meetings or social events away from your board table, excluding the full board. Board business can soon become a "limited mutual admiration society process" away from the necessary transparency and critical analysis provided at the board table. The accepted formalities of the boardroom such as agendas, formal rules and procedures, minutes and tabled documents are utilized for good reason. They attempt to ensure an inclusive and democratic process, not "sidebar governance." It has been said that information is power and it is the basis of our democracy. If your Board Chair or CEO tells the board on some issue: "Don't worry about that,

I will be talking to the minister of government or some other powerful executive at coffee or at a banquet this weekend, so we will work this out", you know your board has been hijacked by the "influential elite."

4. Beware of the "Masters of Business Administration (MBA) Myth" which would like you to believe that anyone with an MBA can solve all your problems with a traditional business solution. That solution almost always means cutting people and services to reduce costs with no concrete plan to raise revenues. Beware of self-serving, too eager to please consultants, who promise to save you money by at least the amount of their fees. They will ask you for your watch and, for a fee, tell you what time it says.

5. Too collegial of a board culture may lull members into not asking the hard questions. Boards can suffer from a risk aversion mentality, which avoids the healthy and creative tension necessary to generate new ideas and new policies from different perspectives. There can be a temptation to not actively question management and allow them to obscure developing problems. Apathy and inaction will undermine your effectiveness and your credibility. It is necessary to confront the reality of the brutal facts. Members should avoid the strategy of demonizing the critical thinker and welcome the valuable contrarian. You need to have one or more individuals on your board with a keen sense of smell and the courage to yell "Horse-poop."

6. A healthy board needs to have a constructive and complementary division of labor among board members. Not all members are qualified or interested in all the same issue areas that boards will encounter. Some will show more expertise and competence in financial matters, others in personnel matters, others in program or project matters, others in advocacy matters and so forth. While there needs to be a general interest and competence on all matters, there is direct benefit in focusing members on issues they are best qualified to handle and most interested in doing. For example, some folks may be more interested in participating in the

regional, provincial, national or international arena. Others may be more informed and interested in cultivating working relationships and partnerships at the local level. A few may hold a particular interest in researching various subjects to inform the board. Yet others may excel in recruitment and professional development matters. Some people on the board are best suited to accept the role of chair or vice-chair.

7. Beware of "Ambitious Too Important Board Members" who see their position as just a steppingstone on a power trip to gain the exposure necessary to seek and obtain higher office. This scenario presents the risk that members will take their eye off the local board governance mandate, to pursue their alternative agenda. It can put members into a state of personal conflict of interests, accepting excessive perks and personal largesse for self-promotion. Beware of developing cliques or subgroups and understand that your board is not a social club, but rather, a business organization. Also, beware of the person who has nothing to say, but keeps saying it and saying it and saying it. Beware of board members who feel the selection of non-regularly scheduled meeting dates and times are totally dependent on their own personal calendar and nobody else's.

8. Beware of "The Christmas Tree Theory." By this I suggest that as some people ascend toward the top of an organizational structure, such as becoming chair of the board, they change their attitude and reduce their effectiveness as a board member. They are ascending from the broader base of the organizational tree up to the smaller more exclusive top of the tree, nearer to the customary angel. I submit that the further those members climb, the more they try to emulate that angel shining at the top. As your top executive officers start to share the podium at all kinds of meetings, with all kinds of important people, while staying and partying at the same hotels, and eating in the same restaurants, they can soon become socialized into accepting the interests of that inside culture. Learning from and courteously cooperating with that informal or formal elite culture is important, but it is harmful when the aspirations of that social

club conflict with the strategic intentions of the board they represent. The angel at the top can become too ceremonial and fail to take the often not so harmonious business positions, necessary to represent the people who sent them to get the job done. There is the danger that these folks at the top will capitulate to inclusive self-interests, placing undue emphasis on re-election and sticking with a clique, not taking the solid stance required to accomplish what they came for. It is important to be respectful and social but not lose track of your mission. Sometimes you can be too nice.

9. Beware of "Media Hypnotism" where folks begin to let the media set the agenda. At some point, the media (newspapers, radio, TV, social) will attempt to tell you where to go to be heard and will sometimes try to lure you into issues that you are not prepared for, or that are not relevant to board business. Controversy sells. Communication, transparency and formal reporting all have crucial places in board and executive life. However as a responsible board member you must mutually consent to a process you feel would be most beneficial to the organization you represent. I must confess that this requirement of self-discipline with respect to the media is very difficult for me to comply with. As a bit of a country bumpkin at times, I find it impossible to not be somewhat smitten by the mythology and power of the media.

10. It is the board's task to turn information they are provided with into practical knowledge. You will need to be data-savvy. Avoid "walk–in information" that arrives at the meeting without any previous opportunity for consideration in preparation for the meeting. Also remember that opinions are not facts. People can develop and defend their own opinions but they cannot invent their own facts. Truth in its simplest form is a fact. Something may "sound correct" but is it valid and reliable information? Recognize how some catchy words like modernization, reform, freedom, engagement, lean, consultation and consensus are often used by the "spinsters" to decorate a completely unrelated or opposite context. Remember: "They spin to win." In this information age be aware of the fashion to overuse words like data-driven, evidence based, standardized

tests or master plans and use them to encourage procrastination when something needs to be done, even if the data or evidence is not all collected. A board can be trapped into a situation of "paralysis by analysis." "It is important to grow the pig, not just weigh the pig."

11. Your political affiliation or ideological preferences cannot usually be hidden from your colleagues, and are best exposed and recognized by all members at the table. Political objectives that conflict with organizational objectives can then be challenged. People who say you can leave your political skin at the door have escaped from reality.

12. Board members should not let themselves become dependent on the compensation and benefits paid for serving on the board. Self-preservation on boards for financial reward is not healthy for productive and independent decision-making. On the other hand, members need to be fairly compensated for the job they do. There needs to be an incentive for recruitment in order to get the best.

13. Board decisions require a majority of the members to agree, since the official power does not rest with the individual. Therefore it is essential to develop and practice your skills as an "influencer." Knowledge, preparation, and respect for others will pay you back in better decision-making.

14. A confident and mature board will accept failure as part of the process and take responsibility for their mistakes, knowing that their work has a moral purpose and is partially directed from the heart. The difference between a professional and a rookie is that the professional has failed many more times.

15. Limit the participation of management and officials to their role as resource people and advisors not final decision makers at the board table. Keep in mind that they work for the board, not the other way around. Hold regularly scheduled in-camera sessions to discuss certain sensitive issues away from management and staff.

16. Good management makes good unions and good unions make good management. There needs to be an environment of mutual respect and common purpose between these groups with the best interests of the organization in mind. You will need to live with each other after an agreement is struck.

17. Note that principles, values, beliefs and commitment are very important human characteristics and that "hypocrisy is the enemy of democracy." The "dumbing-down of ideas" and inventing ridiculous or contradictory compromises to reach a limited consensus is frustrating and makes democracy messy. Remember: Governance is the art of the possible, not just a vision of perfection. However, at some point the elimination of substance for the sake of compromise should reach a walk away point, because the value of a compromise is gone. Stand for something.

18. At your board meetings, include time on the agenda for "round table reports and viewpoints" in order to encourage everyone on the board to have their say and comment on any issue of special interest to them. This is a process that works well to balance discussion time among the more boisterous forceful extroverts and the shy introverts. Set aside sufficient time for organized but informal "BS-sessions"—brainstorming sessions or blue-skying sessions with the assurance that there are no bad ideas and everyone has a turn to present his or her views. Eliminate the "killer comments" like—"This has never been done before"—"We don't have the budget"—"This is not practical" or "This is not what government wants." Make sure you have some idea people mixed with some implementation people involved in the process.

19. It is always important to recognize employee achievements and it is always wise to get to know and take time to visit with the janitor and other support staff that know the true workings of the organization. It is beneficial to adopt high expectations for your organization, because it is better to set high goals and come up a little short, than set your expectations low and meet them.

20. Sometimes it is useful to ask yourself what your mother would decide in these complex circumstances. Plain old common-sense decision-making never goes out of style. The "test of a reasonable person" is your best guidepost. In education circles they call it "the test of a fair and judicious parent."

If you go in the kitchen at some point you will feel the heat. Nobody is perfect and some people always like to remind us. As board members, we live in a house without curtains. One way or the other, we are accountable for our decisions to someone. The life of a board member extends far beyond the board table and the sound of the chairperson's gavel. It can find you deep in the bowels of your organization evaluating the shaky foundation of an aging building, or you may be on a local field trip to see a new project or meet new people in the organization, or you may be meeting the public to represent the board, or you may be at a recognition banquet for employees and management. Further away, you may be in a car or airplane heading to a provincial, national or international meeting or convention. Then there are the hotel rooms and restaurant meals that soon become routine rather than a special pleasure. The trips to the bar and late night rendezvous up in the rooms become part of the lifestyle. The temptation for excessive pleasures is all around. Self-discipline and a moral compass are essentials in order to sustain this lifestyle and maintain your home life.

Remember that as a board member, your performance review is not just a thirty-second media clip, a four-column newspaper article, a two hour public meeting, a three hour written exam, a week long planning lock-up or a fancy retirement party. Your legacy can be all of those things but more importantly it will be the actual results you accomplished, in the short-term crisis mode, in the medium term remedial mode, or in the future planning mode. It will also be what builds or grows from the foundation and inspiration you created. You will get your final marks from your board governance exam when you drive by or visit a magnificent facility you worked in or on, when someone you guided or assisted makes a meaningful accomplishment, when someone mentions how much you helped them succeed, or when you hear someone repeat something you once said and know it is now also their opinion or belief. You will be astonished and proud when you retire and hear genuine praises from people who never really liked you. As difficult as it may be to accept, the pinnacle of flattery is when someone else, often a pompous politician, claims total credit for results you and your board have accomplished. Leadership truly means being a faithful servant for the people and the organization you represent.

Martin Luther King said: "Our lives begin to end the day we become silent about things that matter." Winston Churchill once said: "You have enemies? Good, you have stood up for something sometime in your life."

As a young person, I always wondered why somebody didn't do something about an irritating or senseless issue or problem I observed, but then I realized I was that somebody. I came to realize that we should not adopt a comfortable mindset of powerlessness, because I believe any institution or service built by people, can be changed, destroyed, restored or improved by people just like you and me. This is not a call for rudeness or violence but it is a call for action. We need to rediscover our purpose and develop the courage to commit ourselves to public awareness and citizen participation. Yes we may need to surrender some small amount of our rugged individualism, to collectively achieve the desired results, but it is worth it. We can all make a difference.

Democracy

by Dan Danielson

So everything in life doesn't just go tickity-boo
And our governance processes don't please you

You say that if somebody fixes the democratic state
Then you might just be quite willing to participate

So when we figure out what you want us to do
We actually learn that somebody should be you

If you leave decisions for everyone else to make
You should not be disappointed by any mistake

Poverty, illness, starvation, and wars are insanity
Caring, sharing and helping are common humanity

Our contemporary adventure into globalization
Should not tempt us to betray our home nation

Indifference, apathy, ignorance and personal greed
Will not create the caring world we all really need

If we can take the time to help design our future
Then our kids and grandkids will be more secure

eight

A GOOSE HUNT AT GOOSE RIVER LODGE

I will always remember that last hectic fall day, ending our annual harvest. We are captured in the bright prairie heat, facing a gentle drying breeze, under an immense blue sky, featuring some fluffy white clouds and many irregular shaped flocks of transient geese, ducks, cranes or swans. I can still smell the crisp dry aroma of grain dust particles, from the generous meter-wide swath of pre-cut straw with full heads of durum wheat that feed into the hungry combines.

The front mounted "pick-up" consisting of a horizontal rotary assembly lifts the swath up and by the forward motion of the combine, pulls it into a large crossways "auger" on the table of the machine. That auger with complex spikes or fingers that move in and out, pushes the still unthreshed grain into the enclosed "feeder house." Through the feeder house it is elevated by a chain or paddle assembly up to the rapidly rotating "cylinder." That spinning cylinder with rigid exterior "rub-bars" threshes the grain against a solid metal "concave" with small spaces within its metal rod construction. The straw is forced upward onto an arrangement of "straw walkers" that by vigorously shaking carries the threshed straw out the upper rear of the machine, through a vicious "straw chopper" spreading it back onto the ground. The partially threshed grain kernels have been forced through the concave down onto a top "sieve" that shakes and cleans more of the straw and chaff away from the seed. That process is repeated as the grain falls down to a more

The Mighty Combines *(photo courtesy of Cory and Lynne McEvoy)*

refined sieve. A large cross-mounted "fan" under the combine blows air up into the sieves to facilitate that cleaning process. Once through the sieves, the clean grain falls down onto a "shaker" sheet of metal that brings it further down to an "internal auger," ready to elevate the clean threshed grain up into the storage "hopper." Then it is finally augered into the farm truck, ready to be stored, sold and shipped to a world market. I can still hear the powerful yet soothing roar, clatter and clang of that giant yellow combine as it slowly but surely follows its predetermined path of square rounds harvesting the last huge golden field for another year. Now the whole countryside looks like it has just been at the barbershop and received a buzz cut.

I must admit that I really love the autumn season. There is something quite comforting about moving away from the extreme heat of the summer, and the busiest work season, knowing that there is now some free-time for relaxation, before the harsh cold of winter sets in. You can feel the annual relief in the voices and laughs of farmers, who are finishing the harvest and now have their crops safely stored in large round steel bins. The colours have begun to change from solid shades of green, to earthy yellows and oranges as the leaves prepare to tumble onto the mature tan and grey grasses, next to the dark grey soil below. Cool damp mornings, often with clouds of fog in the air, and a thin layer of white frost on the shingles, give way to a nice warm but short afternoon, before the crispy cold night arrives. The night sky appears clear and close, with stars twinkling in their permanent positions relative to the North Star, the Little Dipper, the Big Dipper and Orion. Under the protection of the night, out of sight high in the sky, you can hear the rhythmic honking, of the Canada Geese moving toward our river from their summer nesting grounds of the north, ready to temporarily feed on our croplands, in preparation for their later trip south for the winter.

Our farmyard, known in hunting season as "Goose River Lodge," is located about three miles north of the South Saskatchewan River, just where the watershed from the backing up of the river from the Gardiner Dam near Outlook, turns the river into Diefenbaker Lake. From a point straight south of our yard at the river looking left or east it is now lake, then looking to the right or west there is a main river flow with a channel that separates and then rejoins the main river, forming an island. Along that island and at different locations in the main-river, there are a few large sandbars sticking up out of the water. Many animals such as deer, moose, coyotes, foxes, and cougars live and hide on the island, in the tall willows and other types of bush. In order for them to feed or hunt in the farmyards and fields, they need to swim across the channel. The sandbars are home for the geese, cranes, and ducks during the nights before they take off at sunrise in dark noisy clouds heading for the fields to feed. By noon or just before, the birds have completed their feeding on the wheat, durum and pea fields, so they return to the river sandbars to again drink and relax. Then about four-thirty in the afternoon they repeat that whole cycle over again, honking, cackling and quacking back to the silent fields to feed before returning to the sandbars at dark for the night.

During any late fall day the sky comes alive with the sounds of geese, cranes, ducks, crows and magpies. You can see the artistic designs created in the sky by the profiles of many, many, flocks of birds. The gangly grey Sandhill Cranes which only weigh up to six or seven pounds, but stretch up to four feet in height, sing a lilting but shrill refrain as they float in converging circles high in the sky, almost out of sight as the warm air currents beneath their wings lift them up and up. They are able to maintain their altitude with only a few gentle flaps of their large wings, huge in relation to their lean, long neck and slight body. They fly effortlessly with their long black legs hanging straight out behind. A pampered cousin of the Sandhill Crane is the larger Whooping Crane, which is almost all white with black tips on its wings, and a dashing red patch on its head. They are a protected bird because of their low numbers and the worry of extinction.

The majestic Canada Goose can weigh up to eighteen pounds with dark backs, a grey chest, long black legs, and nice white patches on its cheeks. When they fly, they flap their wings quite rapidly relative to the gliding cranes, and they tuck their long legs up against their body unlike what a crane would do. The smaller Canada Geese are known as Lesser Canadas. Snow Geese, which are all white except for black tips on their wings, are the most numerous of the geese in Saskatchewan. They have grown to such large numbers that they have become a real nuisance in the fields and sloughs. Hunters are allowed to harvest almost unlimited amounts of white geese—up to twenty per day. Then there are the White Fronts' or "Speckle Bellies" as they are often referred to. Those geese are smaller than the Canada Geese and look more like a Mallard Duck with orange legs. They have a lighter grey chest when younger, until they mature, developing an almost white chest, with three or four blackish bars across the front. Some people describe those bars as speckles. Hence

Mates for Life *(photo courtesy of grant@dougallphotography.com)*

the name: "Speckle Belly." The "Speckle Belly" or White Front in my opinion is the best for roasting, filled with dressing and set with all the trimmings.

Both cranes and geese usually fly in flocks organized in an uneven "V" formation, with one bird as the leader at the apex of the "V" and a number of birds in rows on each side going back to form arms of the "V." Usually there is a different number of birds on each side of the "V." The scientific explanation for this common flying formation is physics based and purports that as each bird flaps its wings it creates an uplift for the bird immediately behind it, making it easier and less tiring for each bird to fly in the row. It is estimated that, in this labour-saving "V," a flock of birds can increase its daily flying distance by seventy percent more than flying alone. Some of my hunting buddies dispute the science and suggest this formation was simply adopted because the front bird has the road map.

The fact that our farm is so close to the river, but far enough away to be past the uncultivated river hills called the Sandhills, makes it ideally located to become over two thousand acres of bird restaurant. There is also one large slough where some of the birds can choose to stay overnight, not wanting to return to the river. As a result, over the years our home on the farm has become a valuable location for enthusiastic bird watchers and hunters. For these reasons my family has lovingly labelled our farmyard "Goose River Lodge." Over the years that place has been visited by many hunters from several places in Canada and the USA. Happy hunters tell us that our location allows them to stay close to the hunt, and with the large slough near the house, the sounds and sights of all kind of birds, day and night, are special attractions, which add to their back to nature experience.

From the comfort and scenery of the "Goose River Lodge," hunters can have three unique migratory bird hunting experiences. They can do what we call "Hill Shooting" or "Pass Shooting" down on the river hills as the birds come over on their way to the fields in the morning. They can hide on the edge of the slough and shoot at the birds as they come or go from the water in the slough. We call that "Slough Shooting." A third choice, which has been very popular over the years, is "Field Shooting," when hunters put up decoys in the field to attract the birds while they hide nearby in blinds or pits. Throughout local history all three of those hunting styles have been enhanced to a precision science by dedicated and creative hunters.

Hill Shooting

Ever since our boys Danton and Brandon were quite young, before they could have a hunting license themselves, we liked to go hill shooting together. I was the only one with a gun, but they learned to shoot with my gun under my close supervision. They loved running after the downed geese or cranes that we managed to shoot. For us, like many farm families, hunting out in the fresh air close to our environment was a delightful bonding experience, often done on our own land or on a close neighbour's ground.

If we wanted to go hill shooting, we would always plan our hunt a few days ahead by calling the rancher who had the pasture land on the river hills where we wanted to hunt, in order to get permission. Then, on hunting day, we would get up about an hour and half before sunrise, get our long underwear, heavy winter coats and hats on, grab the shotgun I kept in a heavy cloth and leather, zipped-up gun case, take a couple boxes of twenty-five shotgun shells, and load up in the warming truck ready to go hunting. We would either go straight south of our yard, to the hills near Weber Spring, where most people hauled their water for household use, or else we would head over to the river hills by Larson's Grove, about two and one-half miles west. On our way, we would always see three or four deer jumping the fence beside the road just ahead of us in our truck lights. We would almost always see the shiny eyes of other deer, standing in the field in the dark. On occasion a deer would stand still in the middle of the road, stunned to stillness in our bright headlights. We would wait for it to regain its awareness and move back into the ditch.

Arriving near the grass-covered hills, the first thing we needed to do was find a place just back of the hilltops to park and hide our truck so as to not spook the birds. Together, we then would walk up the back of the hills on the opposite side from the river. Once at the top of the hill, we would find a nice bluff of small bushes, which we could hide in. Sometimes there would be a small pit, which somebody else had dug before, and was ready made for us to hide in. Often we brought a shovel to enlarge our pit. The boys would run and get some tumbleweeds, or fallen branches, to help us build our hunting fort. The thin prairie-wool grasses we settled into on the ground were moist with the morning dew.

Pretty soon the morning sky in the east would start to get orange red, as the sun would be coming up in about a half an hour. Legally you could shoot birds one-half hour before sunrise, so you needed to be ready if any individual birds came by before the main flocks came out. We used to call those early loners the scouts who we imagined were sent out to look for hunters. Therefore you didn't want to let them get back with the news of our location. The morning air always gets its coldest just as the sun is rising. As we looked down from our hunting blind onto the river, we could see a thick layer of fog, which obscured our vision of any early birds. However, we could hear the amplification of the music from the choir of geese and cranes as they got anxious to lift into the sky. Usually, they would lift off in a series of smaller flocks, which would give us a longer hunt as they all filed by. Not wanting to make it too easy for the hunters, and with their natural instincts for survival, the birds would often fly up or down the river against the wind to gain altitude, before they came over the hills where we were hiding and waiting.

Usually a lone "Speckle Belly" would be the first bird to come right at us, so I would take a shot with my trusty, semi-automatic, Remington twelve-gauge shotgun, shooting three-inch shells with number two steel shot. If you were there you might hear "Darn I missed—maybe I didn't lead that goose enough" meaning I should aim a bit ahead of it, so as to make up for its flying speed. There is a bit of vector physics involved in this goose shooting. "Maybe I led it too much and shot ahead of it." I would ponder. My sense of humour reminded me of the story about three statisticians who went deer hunting. They were going down the trail in a truck when a big buck ran up and stopped right dead centre in front of them. The guy on the left side jumped out and shot his rifle, but shot three-feet behind and missed the deer. Seeing that miss, the driver jumped out, but he shot three-feet in front, also missing the deer. Just then, the third and smartest statistician, sitting in the middle, jumped out with his hunting knife ready to butcher the deer, because he calculated that on average they had killed the deer. That big buck with trophy antlers slowly ran away with a sly smile on his face. Now back to the goose hunt.

As the geese and cranes lifted off the river, they made a chorus of loud melodies, out of shrill and joyous honks and cackles. They would be coming right toward us and some would be right over our hiding spot. I shot, and the boys as they got older also used my gun under my supervision to take a shot or two. Together we would usually quickly knock down five birds, which were our legal limit. We could have no more than two white fronts (Speckle Bellies), out of our five-bird bag limit. The other three at least would need to be Canada Geese or Whites. The boys ran like well-trained hunting dogs to retrieve our game. Our adrenalin levels ran high as this super-happy country dad inhaled the joy my boys showed, as we spent the early mornings outside connected at the heart, with each other and our environment. It was soon time to go home and brag to mom about our successful hunt. Then we would go into the warm shop to pluck and clean our geese, which we loved to eat after Sharon/Mom roasted them in her special way.

When we were home following our hunt and a huge farm breakfast, I usually snuck into my easy chair to catch a few winks, before I challenged the rest of the day. Before I dozed off, my mind often wandered back to a time when my dad, who had not hunted for many years before, suddenly got back into it, when my brother, Eldon and I as well as a few friends needed an adult to go with us to legally hunt birds. For a number of years, my dad got up early in the morning to act as our guide, our father and friend. We loved to hunt together. Those were great times and even long after his passing, my friends still comment about how much they respected my dad, Urvan Danielson for what he did for them, and for whom he was as a person and role model.

Field Shooting

Field Shooting is a second style of hunting for Geese, which can become a special occasion as hunters put together their imaginative and ingenious minds to fool the majestic but sly Canada Goose. This hunting process mixes much work and much pleasure in equal portions. These field-hunting missions are driven by a passion for success, and they also involve planning, innovation, resources, management and evaluation. It is very important for goose hunters, like project workers, to be with the right people, at the right time, for the right reasons. Being friends and liking to hunt birds is not enough. A hunting group needs to trust each other, because no stupidity or carelessness is allowed when it comes to guns. There can be no second chance with guns. Also, members of the group must be ready to do their share of the work that goes with successful hunting.

For over twenty years, I was part of a group of four compatible and happy goose hunters. When I say goose-hunting, that is our generic term meant to include crane hunting, duck hunting, upland game hunting and pheasant hunting, when those seasons are open. Our twenty-some year hunting association always took place on the long Thanksgiving Weekend about October 10th each year, at our friendly "Goose River Lodge." Our hunting party was made up of Brian—a teacher from Lloydminster, Ralph—a businessman from Saskatoon, Darrell—a teacher and Mayor from Melfort and myself—a "suitcase farmer" from Tyner and Saskatoon. My preliminary work preparing for the hunt would be to have the lodge ready, the decoys and pit boards organized, and then "spot the geese." That spotting process I would perform was intended to predict which field the geese would most likely come to feed on in the morning. We simply followed the birds off of the river, carefully watching where they came to feed the evening before our hunt. That was the location we expected them to return to in the morning, although that was not always a fool-proof method for prediction. Significant changes in weather could divert the birds to other places.

Usually, my hunting buddies could get away early on the Friday (I would tease them about having easy jobs) and arrive before the sun went down. On arrival, since we usually only saw each other once a year, we would want to have a drink and quickly get reacquainted. That social function would usually involve getting the binoculars out and driving around our fields again to see where the geese had come out to feed that evening. More than likely on that initial local tour, we would run into neighbours or other hunters from many places, whom our group had come to know over the years. They would be out doing exactly what we were doing—"spotting geese."

Soon our morning hunting spot would be chosen and we would continue to socialize until the birds headed back to the river for the night. At that time the uneasy question of deciding where in the field our "pits" or "blinds" should be located would come up. Our "pits" would be holes dug into the ground for us to hide in. In our early years these pits would be dug by hand as a brutal chore conducted with pick and shovel, in the hard clay ground. Since there were four of us we would pair up and share the work by digging two "double pits." Each pair or team would carefully clear the straw at our chosen field location, in about a six-foot by three-foot rectangle, saving the raked up straw to help conceal the final hole. We would dig down about three feet, maybe a little more, but deep enough so we could sit in our pit, side by side, without our heads being exposed. Then we would have a

Dani, Brynlee, Danton, Dan, Sharon, Karen and Brandon Danielson

Dad and Mom—Urvan and Geneva Danielson

The Saskatchewan Wheat Pool elevator *(photo courtesy of grant@dougallphotography.com)*

The Dan Danielson farmyard

The Saskatoon Public School Division office *(photo courtesy of grant@dougallphotography.com)*

Saskatoon Public School Board Trustees. *Standing:* Bronwyn Eyre, Donna Banks, Darrell Utley, Kevin Waugh, Vernon Linklater, Dan Danielson, Kathleen Brannen; *sitting:* Holly Kelleher, Colleen MacPherson, Ray Morrison *(photo courtesy of the Saskatoon Public School Division)*

Top left: Horse on the prairie; *top right:* "The Meadowlark" *(artist print courtesy of Paulett Sapergia, chickadeedesigns.ca)*; *bottom:* Pronghorn Antelope *(photo courtesy of Glenda Rossignol)*

Top left and middle: Baby White Tail Deer living in the Danielson yard *(photo courtesy of Jack Clary); top right:* Mule Deer doe living in the Danielson yard *(photo courtesy of Grant Dougall, grant@dougallphotography.com); bottom:* the first moose to visit the new Danielson yard.

Cranes in flight *(photo courtesy of Jack Clary, Lancer, Sask.)*

Geese in flight *(photo courtesy Grant Dougall, grant@dougallphotography.com)*

Swans on the slough by the Danielson yard *(photo courtesy of Jack Clary)*

Entering Goose River Lodge

Old Mother Goose and goslings in the slough at Goose River Lodge *(photo courtesy Grant Dougall, grant@dougallphotography.com)*

Sunrise over Goose River Lodge with Geese and Cranes lifting off the slough *(photo courtesy of Jack Clary)*

Dan Danielson and Rod Lee with their home made pit digger

Daryl Halliwell and Dan Danielson after a successful goose hunt—Food For Winter

piece of old binder or swather-canvas with slats that reached across the hole, to hold the canvas over the top. It covered over our heads as we sat in the pit. The straw saved from originally clearing the area for the pits, along with other straw, is strategically placed on and around the pits, to keep them as obscured as possible. "This is a precise science, you know."

The objective of our rather vigorous hiding ritual was to lure unsuspecting geese into our "decoys." Those are fake geese made very lifelike from hard rubber or a moulded cardboard material, which we set up around the pits. Because it is known that birds, like airplanes, try to take off and land against the wind, the decoys are placed facing into the wind if there is any wind.

All of this "pit preparation" happens the night before, no matter how late it gets, or how tired or booze happy the hunters are. However, the decoys are set out the next morning just before the hunt, so they don't collect white frost, making them appear glossy and unreal. Putting up the decoys in the morning can be a daunting chore, with a couple dozen flat plywood decoys, three or four dozen full-body plastic, rubber, or cardboard decoys, about two dozen white decoys and a couple dozen duck decoys. All of this detail and meticulous work is why it requires a special group of friends who like goose hunting enough to do their part. Everyone needs to help to make the hunt successful for everyone else, without anyone having to do more than a fair share of the work.

Back at the lodge after all is prepared, as long time friends we continue to joke and drink into the early morning, not leaving much time to sleep fast before we get up and go to the field. The alarm rings at a little more than an hour before sunrise. We all get up moaning, groaning, grunting, coughing, spitting and grumbling about why we do this every year, at this stage in our lives. Yet we love it and we do it. Getting up and ready is the test that proves you have the right hunting party. Coffee, treats, guns and shells in hand, we load up the truck and head for the field, which—since we are on my farm—is within ten minutes from the yard.

First we take the truck right up to the pits, dropping off the decoys, guns, boxes of shells, etc. using the headlights and interior lights from the truck, because it is still very dark. Then we start to set everything up. Once that job is partially complete, someone takes the truck off the field to hide by a nearby granary or in a ditch by the road, and then walks back to take his place in the pit for the hunt. It would soon be time for the birds to arrive, lured by the realistic and tempting decoys, and the seductive mating call of a mouth blown "goose call." My confidence in using the goose call was shattered for life one time, when some hunters who had set up in an adjacent field, yelled at me to suggest that I should shove that goose call up my ass. I have never completely recovered. Not from the insertion of a goose call but rather from the ridicule to my ego.

The Pit Digger

We had grown weary with hand digging goose pits, using a pick and shovel, so modern technology was about to hit the sport of goose hunting at Goose River Lodge. A couple of neighbours and I decided to try our hand at making a "pit digger." I had my dad's old 650 International Harvester Tractor, which I contributed as the power unit, from which to begin our amateur manufacturing process. Then we found a rear-end differential gear box originally used to drive the wheels of a self-propelled combine. It was designed to work in a horizontal position with wheels on the axles coming out on each side of the differential housing. We welded the axle assembly on the one side locked, and removed the rest of the axle itself, in effect stopping the motion on one side of the differential. Then we turned it from a horizontal to vertical assembly with only the bottom side of the assembly, which we had not locked, still able to move. That assembly would serve as our drilling unit heading down toward the ground.

Then we designed and manufactured a solid angle iron lifting structure, coming off of the hitch drawbar on the back of the tractor, which could be mechanically lifted up and down by a large hydraulic ram. The new structure was about two feet wide and eight feet long, angling up from the tractor drawbar at about a fifty degree angle. The differential assembly hung on the upper end of the new structure, with its remaining live or turning axle facing toward the ground. The hydraulic ram powered by the hydraulic pump on the tractor was adjusted to easily and powerfully lower or lift the whole angle iron assembly. Then we attached a heavy-duty

Power-Takeoff (PTO) drive from the PTO on the tractor to power the turning axle. The drive shaft was telescopic so it could slide to become shorter as the ram pulled the whole unit down, or could extend as the ram pushed the unit up raising or lowering the differential drive up or down from the ground level.

At that point, we needed some type of digging device attached to the down side of the vertical differential drive, which would be sharp enough to dig into the ground as it was turned by the differential powered by the tractor through the PTO. For that we decided to use two sizes of drill stem pipe we got from a local oil drilling company. We solidly attached a short one-foot piece of the slightly larger drill stem pipe to the bottom of the differential unit. Then we attached a longer six-foot piece of a slightly smaller diameter drill stem pipe into that shorter but larger diameter pipe with a shear pin through both pipes. That whole pipe assembly would now turn, driven from the tractor.

The last part of our invention process involved deciding on what would be used to cut through the dirt at ground level at the end of the drilling pipe. For that we agreed to use a very solid half-inch piece of flat sheet metal, which we heated and bent into what looked like a short piece of auger. Then, we welded a thirty-two-inch piece of grader blade, to the bottom lip of our auger. The grader blade was thick and heavy but had a nice sharp edge on the bottom side, which had served its original purpose on a road grader for levelling the surface of our roads. Now our local invention was ready to dig goose pits, which it did with very little trouble for many years. Our cost in terms of material and time was minimal, but the enjoyment we had from that new toy was huge. I guess we proved once again that age-old farm theory that "necessity can be the mother of invention."

I recall the maiden-dig for our new invention as we dug a brand new five-foot deep and thirty-two inch diameter hole, onto which the neighbours' historical two-holer outhouse would be moved in order to get a fresh start. That trial run was a total success and our digger was ready to assume its real purpose as our "Pitdigger," beginning its life long journey from field to field to field, drilling holes for excited and gratified hunters. The size of our pits with our new invention was perfect; you could sit down in them on a five-gallon pail and swivel on your pail to see all directions.

Soon we would also invent a system of pit-boards, that when assembled, would make the pit seem slightly deeper and partially reduce the exposure of the top of the pit. Then, as from our previous hand dug pits, we would use a small binder canvas to cover us up in the pit. Also like before, you would place straw around and on the pit to more effectively hide from the incoming birds. Sometimes the hunter would customize little hand carved shelves inside the pit, to hold his shotgun shells, his coffee and snacks, or maybe his binoculars or goose call. These were pits made for a king, and to us, "Goose Hunting" at "Goose River Lodge" was the life of royalty for a few days each year.

Slough Shooting

Decoys to fool the geese

The third goose hunting style is slough hunting. Slough hunting for geese and ducks adds a third dimension to the sport, as it combines elements of hill shooting or pass shooting, with elements of the field shooting style. Sometimes slough shooting is also referred to as a "back shoot" because it takes place when the birds are on their way back to the slough after morning or evening feeding out in the fields. In essence, the slough hunt usually involves setting up floater decoys somewhere along the edge of the water, next to some land-based decoys just on the shore close to the floaters. The floaters need to be anchored by a strong string and a weight on the slough bottom. Then the hunters hide in the cattails, tall weeds and unharvested crop, which remains close to the edge of the slough. What works really well is when you have two sloughs quite close together, where the geese and ducks want to naturally move back and forth from slough to slough. There are two sloughs, one large one and another medium sized water body, very close to each other on our farm, near the lodge. In addition, there is a small hill which makes one slough hidden from the other, unless the birds gain quite a bit of altitude, which they usually don't have time to do when going from slough to slough. Therefore, the hidden hunters have an element of surprise on their side, when it comes to tricking these wise and observant birds.

The joys of this hunting experience include the beautiful sights of nature from all around, in the sky and in the slough setting. There are the orange, yellow, green, brown, golden, red and grey-blue shades of fall, which cover the surrounding fields, the waterweeds, the bushes and trees, as well as the water and sky. The wind can shift from one direction to another, and grow from a breeze to a roar, within only a few minutes. Rain, mist, snow pellets, and snowflakes can appear unannounced. Temperatures during bird hunting season can range from below freezing to quite comfortable short sleeve levels, from the cold of the morning to the warmest sunny part of the day. As the sun goes down, one never knows how cold it will be. In the hunting world, clouds, rain, wind and fog are friends of the hunter. Dull, rainy conditions make it more difficult for the birds to spot the hunters hidden in the blinds or pits, and wind makes it more difficult for the birds to quickly flare away if they see the hunters. Wearing several layers of warm clothes is the hunter's response to any kind of weather, with the ability to remove or add layers.

Looking across the slough there are patches of waterweeds and cattails, which appear like small islands in a lake. Any trees or bushes around the slough appear quite lonely and are in small numbers on this prairie scene. The water can be clear, flat and shiny or it might be choppy with white caps roaring in the wind. Swimming safely near the middle of the slough, but sometimes venturing closer to the edge, you can see the large majestic white swans, with their glossy white bodies extending high into their large, thick necks. They make a low, melodious, almost bass sound, as they sing their deep self-confident song. Giant long-legged Whooping Cranes swoop by with the lazy, slow, calculated movement of their six-foot wingspans, on their way to an island in the large slough, because they are not really swimming birds. Ducks of all kinds, sizes and shapes, zing back and forth across and just above the water, showing that they are by far the best flyers in the sky, making very little noise as they arrive or depart. Unlike the ducks and often leading to their own demise, the cackling Speckle Belly geese, or the honking Lesser Canada Geese announce their arrival and departure well before they appear. The wise old Canada Geese usually remain very quiet as they arrive and often fool even the veteran hunter. It would not be terribly unusual to see a large black moose or two, standing knee-to-hip-deep in the slough munching on water vegetation. You may even have a curious deer wander up to you as you sit quietly in your hiding place. Ranchers in the area sometimes joke with the hunters, reminding them that a deer doesn't have a bell and a moose doesn't have a saddle.

As an enthusiastic well prepared hunter you manoeuvre your creaking body, your rigid gun, and your extra shells, while hiding in the tall weeds and crop. You attempt to predict where the birds will actually come from, and how you should be lined up for the best deadly shot. You ask yourself if the birds will be coming from the other slough, will they be coming from the river, will they be coming from the fields after feeding, or will they just be crossing the slough from the other side to join your decoys near where you are hiding? The answer usually is "all of the above" or "none of the above." Hunting is a gamble.

"Here comes a flock of geese from the fields to the northwest, just about the perfect height to hit with a good shot. Keep coming—keep coming—they really look like they want to come into our decoys. Now they are setting their wings in a rounded down, pointing shape, and coming, as usual against the heavier than normal breeze. With this wind it will be harder for them to change direction if they spot me and try to get away. Here they are—ten or fifteen birds on each side of the V formation, right in front of me, heading towards our lifelike decoys. The chorus of luring goose sounds I am making with my "goose call" seems to be working. It is time to slide off the safety button next to the trigger, on my twelve-gauge Remington shotgun. My gun is ready to fire." One of our hunting party, usually me, is predetermined to "call the hunt" meaning yell out when it is time for everyone to fire away. "Shoot—Shoot—Shoot I yell into the frosty air." Now is the time for all my hunting buddies to quickly sit up, aim their guns just ahead of this flock of beautiful large Canada Geese, and gently squeeze the triggers. Bang—automatic-reload, Bang—automatic-reload, Bang; I fired three shots at three different large honkers. Down they fall right into the middle of our decoys. Not only the three I shot but eight more came down as my hunting buddies also shot at the same flock.

The hunting adrenalin is really flowing. Joy immediately fills the crisp air along with the deafening sound of total chaos as our shots ring out, disturbing hundreds of other birds on the other side of the slough. They are circling in a frantic dark cloud made up of all kinds of water birds. During that utter confusion, several hundreds of geese and ducks ventured past our waiting but lifeless decoys. However, it is time to collect our game and make sure they are dead. The addictive smell of burnt gunpowder hits our frigid nostrils and the warm touch of freshly harvested game birds, enhance this exciting event, out in the comfortable arms of mother nature. Now it is time for my friends and me to go back to the lodge, dress out these birds and add them to our menu of food for winter. When all was said and done, all four of us in our hunting party had each shot our legal limit of five geese. Slough hunting can be fun, fun, fun.

Hunting as a Community Event

The annual "Migratory Bird Hunting Season" has become an economic and social industry in our province. One of the events that has been happening for about sixty-five years, I am told, is the "Ford Beavis Memorial Breakfast" at the site where the little town of Surbiton used to be, west of Outlook. A couple of generations of the Beavis family, who still have the farm near Surbiton, have moved on to do very well as business entrepreneurs in Saskatoon on their bar and real estate ventures. In memory of their dad and grandfather Ford Beavis, they host an annual, free, outdoor breakfast. The whole affair begins at sunrise and carries on most of the morning. Much care has been taken to prepare an extraordinary feast of side bacon, back bacon, sausages, hash browns, scrambled eggs, sunny side or easy over eggs, toast and jams along with juices and coffee. After you fill up your plate, you move over to a large grouping of rustic picnic tables covered with very clean table-cloths. Everything happens out in the open air regardless of the weather conditions, since it is a true hunting event. Not so far away from the tables where you sit down to eat, there is a homemade bar with some hard liquor to warm the soul and get your mouth and mind into gear. I find it a fun time, and several major problems of the world get solved.

The Lacadena Lions Club also holds their hunting celebration each Thanksgiving long-weekend. On Friday night the hunters meet at the Lions Club House for a few drinks and to register for the next day's event. Plans are made about where to go hunting in the morning and anyone who doesn't have a place to go, or friends to go with, can usually get in with somebody. The next day after the routine hunt is over, there is a late breakfast/lunch at twelve noon sharp, when there is a weigh in and prizes like shogun shells, a hunting knife, or gun case are won for the largest Canada Goose, Speckle Belly, Snow Goose, Sandhill Crane or Mallard Duck. My hunting group would take turns letting one of us pick from all our birds to take the biggest ones into the competition and we have won some nice prizes. A great time is always had by all, when some new friends are made, some new stories are told, and the Lions make some money for their charities.

There is also the Lancer Chokecherry Festival, just south across the river from the Goose River Lodge, by way of the Lancer Ferry. That Ferry can carry about four average-sized vehicles at a time, but it seldom has more than one or two. One year we crossed on the Ferry to go south and west of Lancer and Abbey to hunt upland game birds in what they call "The Great Sandhills." That is an interesting lay of the land, with not much for trees, but it has quite thick bush in places and tall sagebrush areas. There are lots of places for the sharp tailed grouse, or prairie chicken as we call them, to hide and flourish.

That particular time we walked, drove and hunted much of the Sandhills and we were quite successful. So later on in the afternoon we decided to stop in at the Abbey Hotel and pub. Some of the folks were really whooping it up that day, with a banjo player, a guitar player, a piano player, and a fiddle virtuoso. A sing along filled the halls with a combination of noise and music. The bottles were shaking on the tables as the music from the front of the bar converted the usually stale, dusty, beer parlour air into perpetual motion as the floor quaked from the many tapping feet holding the beat from this inviting country-style band. I just knew something exciting was going to happen before we got home that night to the Goose River Lodge. That is, if we even made it home.

As one would expect, as we drank more, the drinks went down easier, and we began to feel free from any worldly cares. The perky young gals, with too-low tops and too-high shorts for October, who were waiting our tables and collecting their tips at costly frequent intervals, told us we should join them later over at Lancer for the "Chokecherry Festival." However, we knew we couldn't go because as we told the waitresses, we would need a shower before we could go to such an event after a long day of hunting. Besides we told them we had no clean duds to wear.

More than willing to go the extra mile for us guys, one of the girls said there was a shower on each of the two floors upstairs. We thought about it but didn't take the bait. Time passed rather quickly that day with the music, friendly conversation, and more than any of us needed to drink. Soon we didn't worry much about driving home, because under those hospitable and carefree circumstances, reality is the farthest thing from our joyful minds. The barmaids reminded us again about the Chokecherry Festival. Then somebody from our group asked what time it started? It was about an hour before the the dance. I guess it was me who then asked the barmaids where the towels were for the showers upstairs. They told us in the closets right by the doors next to the shower. About then, four rather happy hunters from north of the river took our turns, two on each hotel floor, getting scrubbed up for this dance in Lancer. That must have been a sight to see as we scrambled around in a complete state of undress to hit the showers. Much to our surprise, with a collection of our ample tips, two of the barmaids had run across to the store earlier and picked up four new pairs of "Fruit of the Loom" under shorts and four nice souvenir Abbey tee-shirts. We asked the girls to close their eyes as they helped us open the packages and find our new underwear. Then we had to put back on our same old jeans because obviously we had not arrived prepared for any dress-up occasion. Anyway, we were now ready to drive over to Lancer, about ten miles away, to celebrate the annual bird hunting Chokecherry Festival with the locals, who seemed to be very much like us. Fortunately they were also dressed in similar attire to ours. We felt right at home.

The event was much like all other country-dances, when folks let their hair down and have a good time. Although, one of my hunting friends quipped, that this was the first dance he had ever been at when the mothers cut in to dance with the daughters. We sure did have a good time, which included a great lunch that only a rural community can offer. However, even in our state of non-concern, we all knew that we needed to catch the ferry back across the river before midnight. It was like if we missed it we would all turn into pumpkins. Therefore we had to be on our way. We gathered ourselves together, thanked our new friends, and headed home to Goose River Lodge.

On the road north down to the ferry, there is a long straight flat section, but then you come to some wicked curves. Suddenly we could see some lights out in the field, quite a ways from the road at the end of the first sharp curve. As we got closer and closer it looked like those lights were from a car way off the road. That is exactly what we saw as we came up to a very bad car accident. Not only could we see a little yellow car upset and badly crushed, way out in the field, with its lights on, but the horn was blowing eerily into the otherwise deafening quiet fall night, far from the happy voices back at the festival. We could not see anyone in the car or near the car. All of us wanted that blasted horn to stop, so we got a large set of water pump pliers from the toolbox we had with us, and ripped the cable off the battery to stop the horn. That was when the air got deathly quiet for a second or two. Right then we heard some desperate sounds, more like gurgling and gasping than breathing. Someone was down in the tall grass with a deathly injury, including a bluntly smashed facial area. There was a stream of blood dripping off his mangled face onto the ground. Our group, in a state of shock, thought for a minute and decided we couldn't move this man right now but we must do something very quickly to help.

Two of us stayed with him, covering him with coats left by the other two who went to call for help. This was before the days of mobile phones in everybody's pocket, so Ralph and Brian left for the nearest farmyard. They went back up the road a couple of miles to a farmhouse where nobody was home, but the door was customarily unlocked. They went in and found the phone on the wall and rang the number for the police, left conveniently on the wall by the telephone. Since the police were in Lancer to oversee the Chokecherry Festival, they were nearby and arrived at the accident scene within about ten minutes of the call. Together we helped them transfer this unconscious but still breathing accident victim onto their body board and load him into their police vehicle. Just before they sped off to the hospital, the one police officer told us to call the Swift Current RCMP the next day, so they could let us know how this accident victim was. I agreed to call.

My hunting buddies and I agreed that it is surprising how fast one can sober-up when you encounter such an emergency. Our hearts were with this unidentified person, who was so mangled up, choking on his own blood, and unconscious. We were so curious how he would make out, but we were realistic and not very hopeful that he could survive such an ordeal. The next day as they had asked, I called the Swift Current RCMP to get a report on this fellow in the accident. The report was as good as could be expected, and they said the doctor felt he was going to fully recover. The RCMP Corporal also said that if we had not been there as quickly as we were to find him, keep him warm and call them, this man would surely have died. We were very happy to hear that hopeful update, and very proud to have been of assistance. We did not know the person and have never met him since. However, to this day none of us have forgotten the Lancer Chokecherry Festival, as both a happy and sad occasion.

Our annual hunting trips out to the "Goose River Lodge" often also had a very distinct international flavour to them. As you may know, migratory bird hunting in Saskatchewan has attracted many hunters from all over the world, but more often than not our hunting visitors have come from the United States. In my personal situation, one group of Americans have had three generations of hunters come to our fields to hunt. They originate from around Dent and Perm, Minnesota. Over the years the granddads, then the sons, and then the grandsons came. These are ordinary dairy and corn farmers as well as county workers and workers in a cheese plant. They were not your robust rich and glamorous movie star types, who you sometimes hear are spotted in our hunting communities. They were just ordinary, good, hard working friends from another nation. These folks, like most Americans, have served in the military around the world, and have many stories to tell. Those my age or older have served due to the two year mandatory draft. They have experience living out in the wild and know how to camp. As a result, over many years they have set up camp in our shop at the lodge, which has a concrete floor, insulated walls and electric heat. They would come in and use the shower in the house when needed and use the outdoor toilet by the shop. Just the luxuries they needed and wanted for a military style home away from home. After the hunts, we got together to have a few drinks and discuss world issues. We learned much about the military mind of the average American and how some had travelled the world. One fellow had served on the US Aircraft Carrier the "Kittyhawk." We also learned the plight of the American dairy farmer and county workers. We always agreed that the Canadian publicly paid medical care system is more comprehensive, transportable, universal and inexpensive than the US private insurance system, if you could even afford to get insurance down there.

In those days, we were not allowed to hunt on Sundays. Therefore on the Sunday of the Thanksgiving weekend we would hold the famous USA versus Canada touch football game. The competition, complete with food, drink and even neighbours who arrived as spectators, would take place in the open yard at the Lodge. Never mind the score, we are all friends at the bar table. At the end of a few days hunting, we would say our manly goodbyes hoping to meet again next year.

Nature Is My Friend

As everyone leaves for home after our annual hunt, I can sit back in my easy chair and contemplate the intrinsic value of nature that we continue to enjoy at the Goose River Lodge, soon to return to its original purpose as the "Danielson Farmyard." I can finally take the time to focus my curious eyes on the colourful autumn trees, that are feeling a gentle autumn breeze as their golden yellow leaves quietly fall, one or two at a time, onto the cool moist ground. The trees are pulling their energy back from the leaves into the branches and trunk, preparing for the predictable harsh winter of cold and reduced sunshine that lies ahead. Hunting out in nature from crispy yellow sunrises to warm orange sunsets, gives me the feeling of intimacy with our environment. I can now take stock on how generous nature has been over the past year since our last hunting experience. Not just by giving us the fruits of our farm labour, but also the excitement and satisfaction gleaned from seeing, hearing, touching or smelling all the living things around us.

I have seen the tender green crops sprout from a seed and force their way through the packed moist soil up into the sunlight of the day. As they stool out into many broad leaves and stalks, they drink the moisture and absorb the sun, growing into strong and secure plants. Soon they are ready to reach the shot blade or bolting stage forming heads, pods or balls producing various flowers for pollination enabling the production of seeds. Those wheat seeds, canary seeds, canola seeds, lentil seeds, flax seeds, pea seeds, or barley seeds become our harvest for the year. I have heard the giant machinery used to grow our bountiful crops. I have felt the morning dew, the gentle rain, the heat of the sun, the fertile soil, the growing plants and even the harvested seeds. I smelled the fresh air, the machinery exhaust, burning rubbish, flowering plants and meals in the field.

I also remember just before our hunt when the garden season ended and mega bags of red Pontiac or white Netted Gem potatoes were harvested from our garden at the Lodge. Sharon still looks at me "crossways" when the waitress in the city asks me what kind of potato I would like with my meal, and I say a Pontiac or a Netted Gem, please. Sharon always grew the largest heads of cabbage you have ever seen, hundreds of multi-sized carrots, ample peas and corn to blanch and freeze, along with beets and turnips. There were tomatoes and cucumbers table ready right from the garden. I can still taste the delicious chokecherry jam or wine made straight from the bushes at our lodge. "Old Faithful," our trusted electric pump, bringing timely irrigation water from our precious dugout to the garden has been a great success story. The proof is in the produce.

Right beside our dugout, balancing on a bursting brown cattail, I could see a beautiful red-winged blackbird tilting back and forth in the gentle wind. Almost as if singing sweet melodies to the blackbird a chunky grey meadowlark with a large black V across her large yellow breast, sits on a corner post at the end of a falling-down fence. These birds seem well aware of their natural beauty and musical talent and don't hesitate to flaunt it. I remember watching with interest as a busy bee floated from plant to colourful and sweet smelling plant to accomplish its life-creating pollination duties. Even a beautiful orange and black monarch butterfly would happen by to enhance my memories. I remember that original young, almost black, bull moose, the very first to arrive in our territory, as he stopped at our new dugout in 1983 for a drink and a rest in our trees. From that day on it is not surprising to see moose in our trees or somewhere nearby.

In my mind's eye I see that little late-season, spotted fawn, hobbling on his new legs as he scurries away into the protective maple trees in our yard. I can see those handsome brown, red and black, ring-neck pheasants that have fully matured since the Kyle Wildlife Federation released them as chicks in the spring down at "Coyote Bluff" by the river. Right near where I saw the pheasants, I spotted the churned-up loose sandy soil where the sharp tailed grouse or "Prairie Chickens" do their crazy high-stepping mating dance. It would remind you of us at the Lancer Chokecherry Festival. I recall the regal Bald Eagle's arrival every October ready to perch on our fifty-foot cottonwood tree, resting from a busy hunt in the fields.

I can still hear the haunting hoot of the multi-coloured pigeons nesting in the corners of the rafters inside an old rundown, sun faded, pale green grain bin. Over time, they have painted everything around that area a ceramic white with pigeon crap. One by one they frantically fly out, escaping into the bright sunny blue sky as you approach their messy home. I remember that regal but lazy old pronghorn antelope, resting right on top of a wheat pile, while three females grazed on sprouting grain around the edges.

Near where we sometimes cleaned the birds out in the taller poplar trees, I remember the high-up nest of four little baby horned owls being fed by their parents on shore birds like small ducks, killdeers etc. hunted from our nearby slough. I remember watching a curious gray coyote and a sly red fox team up in an attempt to scare up hundreds of startled and noisy geese and cranes from a typical eighty-acre harvested wheat field. They are hoping to catch a wounded bird or two that could not lift off the ground. It is the balance of nature. I remember watching a brownish Swainson's Hawk dive down from the hot blue sky to grab a defenceless gray field mouse with its humongous piercing claws. I also recall how those hawks and seagulls would fly low behind a tractor and cultivator in the field, waiting for mice to be chased out from their hiding places. I remember those speedy striped gophers that have survived from the grips of their vicious predators and lived at the lodge for as long as we have been there. I also remember squeaking back at a bold striped gopher that would confidently talk to me as I worked in my shop.

I remember walking among those stocky, white-faced, red, hereford cattle, and petting a stately sorrel-tan horse with white markings on its face and socks. Seeing them, rewound my mind back several years, to memories of some real working ranches, now relocated because where they sat is now completely flooded, since the Gardiner Dam backed up the South Saskatchewan River into the gigantic Diefenbaker Lake. I remember on different carefree occasions, slowly driving our truck with my dad as my driving instructor, stopping at four different ranches along that flat, sandy road, through the river valley. We received coffee or juice, snacks and good old country hospitality. I remember how they invited my whole family back to pick delicious Saskatoon Berries.

I can still hear the childish belly laughter of our two precious boys, as in the fall we fished out some pan-sized rainbow trout, that we had planted as fingerlings, in our huge dugout, in early spring. I can also hear the laughs and yells of fellow hunters, out behind our trees, throwing up clay targets to practice shooting their trusty shotguns.

As I shuffle back to life from that amazing daydream about our experiences in the great outdoors, I can't help but wonder about our environment, and speculate as to what our future will hold. Will we be wise enough to protect all this beautiful scenery, music and production of nature?

"The Meadowlark" *(artist print courtesy of Paulett Sapergia, chickadeedesigns.ca)*

nine

Conclusion: We Can All Make a Difference

I am very proud to say that I grew up in the small rural town of Tyner, Saskatchewan, which came into existence in the mid-1920s along with five other prairie towns placed about eight miles apart, along a CNR branch line. That railway served for many years as a lifeline, threading together our small farm communities like pearls of the prairies. My reason for moving with my parents from my grandparents'1912 homestead farmyard to Tyner at age six in 1955 was to attend school and begin my formal life-long-learning adventure. It was the requirement for education that brought me to my hometown and later it was education that took me away. For the next twelve years of my life I lived in Tyner, a town that offered me a full course of family love and joy as my comfortable country home. Then for a few years I went off to university in the cold of winter, returning to my home area to work on a neighbour's farm for the summer months. That employment opportunity allowed me to maintain my home roots but also get my education. I think it even helped me pick up my share of the world famous "Prairie work ethic."

Soon after completing my first university degree I moved to Regina to work for the government. However before long I found an opportunity to begin establishing my own farm back home. It meant working on the farm and also taking off-farm jobs in the winter to subsidize my farm dream. For most of my life I have been caught up in a rigorous cycle of education, farming and employment. Each component supported the others. My family viewed education as the greatest equalizer in our society, and for me it has expanded my general knowledge and opened many interesting doors of employment. I think you will agree that the term "Suitcase Farmer" accurately describes my somewhat nomadic lifestyle. In order to fit everything into a busy life with my diverse career and family experiences, I would leave my heavy old Samsonite suitcase—the one my wife backed over with the truck—partially packed near my bed, ready to be filled to the top and forcefully closed en route to a different bedroom location close to another job, the university, or a boardroom meeting. That spirit of mobility was best illustrated when I would get off my tractor in the field on our farm, about eleven o'clock in the morning, then travel into Saskatoon to chair the largest urban school board in the province that afternoon and evening.

My trusty old suitcase

I sincerely believe that the dominant motivation for me to leave a good job in the city and start my own family farm was rooted in the irresistible challenge I knew it would present. On more than a few occasions I was politely advised by some of the "wise guys" of the "farm world" that it could not be done, so I often felt like I was doing the impossible.

Prior to my entry into the farm industry, my parents had already lived through the days of the "Progressive Movement" and the "Tommy Douglas CCF Years" which together provided a productive experiment in social democracy. They fondly informed me how farmers and their leaders, professionals and their associations, labourers and their unions, from rural, urban, and northern communities, came together in a spirit of co-operation to share the human, financial, and natural resources needed to build our province. Rural electrification, telephone service, bus service, hospitalization, and medical care insurance are lasting examples of that success. I learned from them that the wealth of our province at that time had a more direct flow to the common people.

For Sharon, my life's partner, and I, growing a new farm and a new family at the same time was an opportunity of our lifetime. We loved what we were doing. However, we will be the first to admit that we got a critical leg-up from the Saskatchewan Land Bank, the Farm Credit Corporation, various co-operatives and other farm assistance programs. In that era when we grew our family farm, we very much respected the spirit of co-operation as demonstrated by the retail co-ops, the credit unions, the Saskatchewan Wheat Pool and the Canadian Wheat Board. We appreciated the security they provided to our farm and our local communities. The principles of local ownership, democratic control and patronage dividends were what we admired. We felt like we had ownership and control of our farming destiny. I recall quite vividly a farm culture of inclusiveness and caring, with a lively community spirit sustained by sharing, socialization, and commitment. New farmers were seen as valuable assets to the community, the farm industry and the country. We felt at home and welcome in the Tyner district.

There are always lessons for us in history. I believe there are many accomplishments from our past that still hold significant value for our future. I also believe that if you make accomplishments in the name of progress but don't nurture and sustain that progress, with education and member relations, it will soon be distorted or lost. Indifference, apathy and personal greed, fuelled by a false comfort that meaningful progress can never be taken away, can be dangerous. The Land Bank, the Crow Freight Rate, the Saskatchewan Wheat Pool and the Canadian Wheat Board are real life examples of such losses. In that context our public medical care system, public education systems, crown corporations, labour unions, national parks, environmental protection laws, and local co-operatives may also be at risk. Moving forward hell-bent on the destruction of our past, making changes only for the sake of change or corporate profit, is not progress, rather it is insanity. Using words like modernization, reform and freedom to mask that insanity only makes life more confusing. In my view learning from the past and preserving previous successes for our future is the essence of wisdom. The practical application of that wisdom will not be easy, because it will require new generations of intelligent, caring and courageous people, willing to learn from history and when appropriate adapt it for the future. Not an easy task.

My dad, if he were here today, would say "Shame on you farmers for letting them take away your co-operatives and your progressive government programs. Why did you stand by and let the Harper Conservative government take away your freedom to sell directly into the world grain market, through the Canadian Wheat Board? You need to stand up for your farms, your communities and our country because if you don't stand for something you will fall for anything." Then he would say "Sure you will meet with some criticism but you can take it, I know you can" and he would quote Aristotle, who said, "There is only one way to avoid criticism and that is to say nothing, do nothing, and be nothing." In my father's day, folks like him in the "Progressive Movement" collectively stood firm against the corporate monopolies and self-serving politicians. I know my dad would want us to restore, modernize, and protect the progressive ideals and institutions his farm neighbours and he worked so hard to gain for us, not for any sentimental value but for the business and community value. He would also say, "You owe it to yourselves."

There is something about a trip to the hospital like the one I made in 2006, as the reluctant guest of honour in an ambulance, with the high pitched siren screaming, while frantic people attached things to me and asked hundreds of questions, that makes a person stop to seriously think about life. I made that frightful voyage while working part-time as a hail adjuster, in late July, for Co-op Hail Insurance. I had just met a farmer in Langham near Saskatoon to find out which canola field I was to evaluate for hail damage. Then I set out for a field location about nine miles from town. He was going to be busy with another adjuster in another field some distance away, so I set out on my own in my truck.

At the field just after nine in the morning I began to pull about twenty-five canola plants to assess the damage and calculate a percentage of loss at that location. I wanted to do a count at all four sides of the field and take an average to represent my field evaluation. As I headed over and arrived at the second count location, I began to feel a hard pressure pain in my upper-central-chest area. Very soon I started to feel weak and realized this was not any pain I had felt before. I wasn't able to get out of the truck and I began to mildly panic as I tried to decide what to do. If I called someone on my cell phone I couldn't be sure how long it would take for someone to find me. With that in mind I decided to attempt to drive about thirty miles back to our home in Saskatoon. As I drove I felt weaker and weaker but I remember telling myself, that if I was going to pass out I should aim my truck to the ditch not at the line of oncoming traffic, where I would endanger someone else.

Fortunately I reached home where two of my neighbours were visiting out in front of our place. When I pulled up and stopped, I desperately and quietly said I had a bad chest pain. One fellow instantly called 911 and the other neighbour got my key from my pants and got me into our house onto the couch. Within around four minutes the ambulance arrived and the paramedics began to work on me. The person in charge, a no-nonsense lady who definitely understood her job said: "Mr. Danielson we are going to Royal University Hospital-you are having a heart attack."

As they wheeled me on the gurney outside to the ambulance and began to load me inside, I remember the young boy and girl from across the street, shyly but aggressively coming near me to see how I was. I tried to be cool as I raised my hand to wave to them as they waved back and I was on my way. On the road the paramedics called the hospital and called Sharon at work so she could meet us when we arrived at the hospital. By twelve-thirty they had put a "stent" or "culvert" as we farmers call it, into my plugged artery and saved my life. Then within a few days they got me back to put in three more "stents" and I have been good ever since, except my wife says there is a difference between being healthy and being good.

I learned from my health event that it is quite natural to face periods of hyper-emotion when you are forced to abruptly contemplate your mortality and life in general. For me I quickly acknowledged how fortunate I was to be surrounded by the love of my family and my friends as well as colleagues and business acquaintances. I was humbled to hear that my colleagues from the Catholic School Board, who because I serve on the Public School Board are often seen as my opponents, took the time at their board table to pray for me. Then I couldn't get it out of my mind how fortunate we are to have a public health care system with the expertise it offers at virtually no cost to the patient. I began to think how happy and appreciative we should be of those folks like Woodrow Lloyd who fought the messy fight to implement our publicly funded, publicly administered, universal, comprehensive and transportable medical care insurance plan. Then I began to contemplate how precious life on this earth is and how we should improve it for our children, grandchildren and their children and grandchildren.

For some reason a few very worldly issues began to fully occupy my mind as if a greater power was saying "Daniel this is your second chance at life, so what are you going to do about helping others face up to some of life's greater challenges?"

I became hypersensitive to the fact that of our world's total population of eight billion people, there are eighty privileged billionaires who combined hold more wealth than four billion other citizens. Put simply eighty rich people are better off than the poorest half of our world's total population. At the same time approximately sixty percent of our Canadian aboriginal population live in poverty. As a society we will need to find a way to more evenly distribute our wealth. In a country as rich as Canada, there should be no child poverty or homelessness. Otherwise we are headed toward class warfare. I agree completely with former US president Jimmy Carter when he states: "If you don't want your tax dollars to help the poor, then stop saying you want a country based on Christian values because you don't." Something is terribly wrong when on January 2 at noon the CEO has already made as much income and benefits, as the average worker in the company will make all year. It is wrong for corporations to pay more in compensation and benefits to its CEO annually, than it does in taxes needed to pay for our public goods and services.

It suddenly became unusually frustrating for me to almost constantly repeat in my mind the question of why religions can conclude that they should torture, behead, or bomb people into accepting their particular beliefs and values. In addition, the thought of bombing and senselessly killing innocent children and families to inflate the profits of the military business establishment was equally sensitized in my whirling mind. I told myself over and over again that no matter how justified it might seem to be, war means someone just like me or a member of my family will pay with life, limb and mind. War is not progress.

I became very concerned about the ineffective nature of our electoral system. Under our first past the post, winner takes all, democratic process, if one of three or four political parties obtains just thirty-five percent of the vote when only sixty percent of eligible voters actually vote, it can gain power. That means that depending on how the votes are distributed across the constituencies in our country, an all-powerful government can take office with only about thirty percent of the popular vote. Simply put the majority of the voters have voted against the government. Unfortunately, when a government is just elected by a minority, it only governs for that minority. In my opinion that is not a democratic mandate. With a form of proportional representation all parties would be entitled to representation in parliament based on their share of the vote. In that scenario all votes would have a greater value in the democratic process. Most other democracies now have some type of proportional representation.

I realized that our society both in the country and in the city has drifted from more of a "we culture" and a "we economy" to a "me culture" and a "me economy." The once prominent philosophy of sharing, co-operation, and government support, resulting in necessary public infrastructure and essential public goods has eroded into a desperate competition of self-promotion for individual and corporate welfare. In my opinion, a healthy rebalancing of that "we-me" relationship is necessary. From the "we" perspective, public goods like clean air, clean water, clean electricity, a safe and secure food supply, pristine parks, new schools, new health care facilities, bridges, roads, etc., will require a huge human, financial and regulatory investment. In turn, that investment can also support the "me" objective for personal and corporate wealth by way of employment and manufacturing. Everyone benefits from the creation of new public goods.

More specifically, I concluded that we need to make the case for preserving and reclaiming the environment, with knowledge, courage, and dignity, explaining that there is much to worry about and much needs to be done immediately. Solar and wind energy infrastructure should be part of our new reality. We can put the "eco" into economics. Without a safe and sustainable environment everything else becomes academic. It seems preposterous to me that we need to fight with our own government and several corporate citizens, in order to save our planet. True happiness, in my view, comes from our contribution to a better society, and a safe world for those who follow us.

I began to understand that, in the greater context, there are not just farm issues or city issues, but rather there are prevailing people issues. I finally knew that whether my suitcase was sitting calmly in the pristine country air at the farm, listening to the laughter of a family working together, or on a sophisticated wing-back chair near an open window in a city hotel, soothed by the sound of happy kids in a park, it was always near the simple virtues of our common prairie life.

As responsible farm and city folks, we need to come together to make some pivotal decisions about where we really want our society to go. The time is right and it is not too late. We can start by acknowledging the real issues, cutting through the spin, improving our knowledge, and beginning to actively participate in the decision-making governance structures. Governments will need to step up to the plate, resist the self-serving corporate lobbies, and exercise their sovereign power to do what is right for all voters in our democracy. As my mom, Geneva Danielson, said repeatedly: "Decisions are made by the people who go to the meetings. However, before you go it will be helpful to inform yourself on the current issues and develop a comprehensive understanding of how board and committee governance actually works. Life can be unfair but that is no reason to give up" Then she would say "It is hard to defeat a good idea-and our future depends on you for those ideas." I believe the greatest obstacles facing our future success are our own indifference, apathy and false greed. I am told Plato once stated: "Some people are afraid of the dark but others are afraid of the light." We should not be afraid of the light, and new beginnings need not erase successes of the past. Although I did not hear him say it myself, I believe it was Einstein who stated, "Those who have the privilege to know have the responsibility to act."

On the lighter side for a moment, there is a story I like to tell about my stay in the hospital. It explains how important it is to speak clearly when you have something to say or a question to ask. This guy in the bed next to me had tubes and wires coming out of him everywhere as he tried to call over a pretty young nurse. He tried to sit up and pull the blanket, wires and tubes aside from his face as he mumbled something like "are my testicles black?" The young nurse was shocked and shyly went out front to the older head nurse at the desk. I heard her tell her boss that the fellow in bed two wants to know if his testicles are black. The senior nurse said, "You know our hospital policy—if a patient asks a question about his health we must answer him-I will go with you to check it out." As they both headed toward my new friend's bed, two other curious young nurses decided to go along for a look. When they got almost back to the guy's bed and before they could look under his covers, he again tried to pull the tubes and wires away from his mouth and firmly said, "Are my test-results back?" So remember to speak clearly when you have something important to say.

Sometimes it can be the way you say something that can make a huge difference. There is that story about a cowboy who rode his horse up to the saloon and swaggered in through the swinging doors and sat at the bar ordering a whiskey. In a few minutes somebody came in from outside and informed the cowboy that his horse was gone.

The cowboy jumped up and went out onto the wooden step. Then he fired his revolver into the air and in a loud deep convincing voice declared: "I don't want to have to do what I did back in Dodge when my horse was stolen" then he further stated "My horse better be back here in five minutes or I will do just what I did in Dodge." The forceful cowboy slowly but confidently went back into the bar and sat down. In about three short minutes his horse was right back in front of the saloon. With a bright smile on his usually stern face, the cowboy headed back outside and got on his trusty steed. Just before he rode away someone yelled; "Hey cowboy, what did you do when they stole your horse back in Dodge?" The cowboy replied; "I didn't want to do it but I walked home."

There is another story going around about when the Minister of Agriculture made a trip out to the Tyner area, soon after his government eliminated the CWB single selling desk, without the required farmers' vote. He came with his entitlement perks of a huge black limousine and a handsome young chauffeur.

Just north of town, by a farmyard, a donkey ran onto the road and because they were driving a little too fast they hit and killed it. The minister true to form not wanting to take the blame for anything, told his driver to walk down to the farmer's yard and explain what he had done. He advised him to use the "spin" that the speed laws don't apply to the minister and that without the donkey the farmer will have more freedom. The chauffeur left the minister in the car for about four hours before he came back. When the driver came whistling back, his hair was messed, his clothes were ruffled and he had a large grin on his youthful round face. The minister by now in a grumpy mood asked what happened to take so long.

The chauffeur said that when he told the farmer what happened he brought out his best jug of home brew and they had a few belts. Then the farmer's wife brought out the best tasting farm made apple pie you could imagine. After that the farmer's daughter asked him upstairs to show her appreciation. That was when the minister jumped in to take credit and observed that his advice on the spin must have generated too much goodwill. The driver replied, "No—I just told them I was the Minister of Agriculture's chauffeur and I ran over the Jack Ass."

I feel very fortunate to have been able to experience both the rural and urban ways of life almost simultaneously over my lifetime. My dual streams of experience have reinforced each other and allowed my family to have the best of both worlds. Some farm folks say they could never live in the city and some city folks say they could never live on the farm. For me I would not have wanted to live only at either, because I wanted to live at both. I believe as a "Suitcase Farmer" not completely immersed in the city or the country there is a certain degree of objectivity when evaluating both cultures from a safe distance outside. What I do know is that we must work together.

Unfortunately much from my past is gone. Even if I decided to pull up my life and move back to my past it would not be there for me. Tyner has virtually disappeared. The two stores and post office are gone, the two rinks are gone, the three elevators are gone, the railroad is gone, the school is gone, the hall has been moved away and the two farms at each end of town are gone. In addition the Chinese café is gone, the Harvest Moon Ball no longer takes place, there are very few sports days, and White Bear Pub has evolved into a popular restaurant. In addition the traditional family farm is quickly disappearing to be replaced with giant corporate farms.

Our tarpaper-thin, all-weather oil road, which optimistically and irresponsibly was expected to absorb the added burden from the elimination of the railroad, has become weary with bumps, potholes and caving edges. Yet without much repair it is desperately attempting to hold what is left of our frail communities together. The social fabric once boisterous and extraverted has shyly withdrawn into segmented spurts of moderate socialization, mainly among family members of the same large farm operation. There are limited options for contact with other neighbours, and sad but true, the modern culture of competition and greed rooted in farm expansion has gradually depleted once friendly relationships. Most of the people from my happy days in Tyner are gone and the remaining folks are confined to relative isolation, unless they venture away to find new friends.

I know we can't go back in time to live in the past, but we can learn from history and go back temporarily for a humourous and romantic visit through the heartfelt "Memories Of A Suitcase Farmer." Memories of milking the cow; skating on the dugout; performing in the school concert; shopping at the general stores; purchasing a pop, large chocolate bar, and small bag of potato chips, together for twenty-five cents; swimming at the river; falling down the Larson's Grove Hill at the graduation party; visiting with the "wise guys" in the White Bear Pub; playing ball at the sports days; building a family farm with the support of the Land Bank; moving our farm house home from twelve miles away and fixing it up; raising a happy family; learning how a combine and grain elevator operate; building a pit digger; participating on a multitude of local and provincial boards and committees; chairing many important meetings; working at interesting off-farm jobs; dressing out a big "speckle belly" goose ready to be roasted for dinner; planning and dancing at the "Harvest Moon Ball"; eating ten pieces of pumpkin pie with real whipped cream at the Tyner curling bonspiel; watching the first moose arrive in our area; catching a glimpse of a majestic pronghorn antelope resting on a wheat pile; observing the white tail doe and the two spotted fawns she raised each year in our trees; squeaking back at a bold striped gopher who talked to me as I worked in the shop; and listening to the sweet sound of a meadowlark singing from a post in our yard.

I now know that on the path from family farm to city boardroom, the country stayed in the boy. Being a farmer is not just what you do because it is also who you are. Thank you for taking this informal literary journey with me, and I hope you had as much fun hearing my stories as I had telling them. In the words of my dad "Urvan Danielson" as he always ended his numerous telephone calls with family and friends, from whom he never wanted to be very far away, "Bye for now."—"Keep in touch."

When we began building our new farmyard, Sharon gave me a Scotch Pine tree for my birthday. It was only four feet in height. We planted it together as a symbol for our future family farm and farmyard growth. Now, our symbolic tree is higher than the farm home.

Acknowledgments

There are a number of people I want to thank for their invaluable advice and assistance as I completed this work. For helping me edit and proofread the many drafts, I owe a special thanks to Harold Pexa, Alec Postnikoff, Darrell Collins, Wayne Welte, Marlow Rafoss, Wayne Corbett, Gene Pura, Margi Corbett, Sandra Corbett and Sharon Danielson.

For the great pictures I owe much thanks to Grant Dougall of Saskatoon (grant@dougallphotography.com), Wayne Corbett of Saskatoon, Glenda Rossignol of Assiniboia, Sask. (Saskatchewan wildlife photographer), Paulett Sapergia (Saskatchewan wildlife artist—chickadeedesigns.ca), Jack Clary of Lancer, Sask. (Saskatchewan Wildlife Photographer), Cory and Lynne McEvoy of Eston, Nancy Stevenson of Eston, Colleen Myers of Saskatoon and *The StarPhoenix* (Saskatoon). The book cover photo is courtesy of *The StarPhoenix* (Saskatoon).

The prominent folks who made a preliminary read of the complete manuscript, and wrote the "reviews" or "blurbs" for the back cover deserve a great big thank you. Thanks to David McGrane, Pat Trask, Alec Postnikoff, Ken Rosaasen, and Wes Funk for doing that very helpful job.

I want to thank Mike McCoy and Brian Smith of Articulate Eye, Saskatoon, for transforming my bare manuscript into a living book, complete with the strategically arranged pictures. Your work is of great quality.

Coffee row has also been a remarkable source of information and guidance as I put this writing together. Thanks to Alec, Marlow, Harold, Harold and Harold, Darrell, Gene, Peter, Oliver, Myron, Andy, Mark, Rusty, Bernie, Steve, Wayne, Neil, Keith, and several others.

For some brutally honest and critical analysis and support, I owe a special thanks to Sharon, Danton and Dani, and Brandon and Karen who are my wife, our two sons and their wives. Last but not least, I want to thank Brynlee and Asher, our grandkids, who without many words have helped me, more than they even know, to think about a future world.

About the Author

Dan Danielson was born at Eston, Saskatchewan on August 12, 1949 and lived on his grandparents' homestead farm for six years prior to his family moving to Tyner, Saskatchewan. He completed grade eight in Tyner before travelling by bus to Eston High School where he graduated in 1967. Then he went on to university in Saskatoon where he completed a B.Ed in teaching, a BA in economics, a PGD (Post Graduate Diploma) in educational administration, and a M.Ed in educational administration. His Masters Thesis was written on policy making and governance. He has a love for "Life Long Learning."

Dan worked with the Saskatchewan Land Bank, the Saskatchewan Farm Ownership Board, the National Farmers Union, the Federal-Provincial Partnership Agreement On Rural Development, and the Meadow Lake School Division. He and his wife Sharon, an employee of Affinity Credit Union in Saskatoon, worked on and off the farm, to establish a successful family farm, build a new home in Saskatoon and purchase a cottage at Christopher Lake.

Dan has served on numerous boards and committees including: the Saskatchewan Wheat Pool Committee, the Crown Land Tenants

Association, the Saskatoon Public School Board, the Saskatchewan School Boards Association, the Saskatchewan Assessment Management Agency, the Primrose Resources Economic Development Corporation, the Committee for a Democratic Co-op, and the Saskatchewan Wheat Development Commission. He has a passion for democratic participation and co-operation that he feels are needed now as much as ever to solve many social and economic problems.

Dan Danielson has a natural flair for the clever and comical side of life, no matter how serious some issues can be. In addition he enjoys the beauty and friendship of nature and still loves to participate in a goose hunt, near his farmyard by Tyner, that he calls "Goose River Lodge." Above all he wants his family and others to have a safe, productive and happy life going into a complex future.